Power Transmission System Analysis Against Faults and Attacks

Power Transmission System Analysis Against Faults and Attacks

Tamalika Chowdhury, Abhijit Chakrabarti, and
Chandan Kumar Chanda

CRC Press
Taylor & Francis Group
Boca Raton London New York

CRC Press is an imprint of the
Taylor & Francis Group, an **informa** business

First edition published 2021

by CRC Press
6000 Broken Sound Parkway NW, Suite 300, Boca Raton, FL 33487-2742

and by CRC Press
2 Park Square, Milton Park, Abingdon, Oxon, OX14 4RN

ISBN: 978-0-367-49777-4 (hbk)
ISBN: 978-0-367-49954-9 (pbk)
ISBN: 978-1-003-04733-9 (ebk)

Typeset in Times
by codeMantra

Visit the CRC Press Web site at https://www.routledge.com/Power-Transmission-System-Analysis-Against-Faults-and-Attacks/Chowdhury-Chakrabarti-Chanda/p/book/9780367497774

To my husband and parents

–Tamalika Chowdhury

To my spouse

–Abhijit Chakrabarti

To my mother and grandson

–Chandan Kumar Chanda

Contents

Foreword

Dynamic and diverse disruptions are continuously challenging the functionality of power grid. Modern communication technologies lead to new hazards of attacks and outages in the past few years, which need to be addressed with the integration of distributed energy resources. This requires the identification of critical buses and lines in power grid networks that need to be protected and handled for vulnerability assessment. The vulnerability assessment process involves mathematical analysis employing different tools, metrics and indices of complex network theory.

Further, it is important that power grid networks recover after an outage to minimise the economic, social and political impacts. Resilience evaluates response and recovery processes. An assessment of resilience performance analysis is attempted considering statistical concepts of percolation threshold that evolved around the response of the grid against any kind of disruption. A framework for assessment of cascading failure has been exploited, which involves preferential attachment model to determine the consequence of grid network islanding.

Modern power system security is prone to unplanned outages due to natural calamity and cyber threats that may lead to destructive consequences for the power system network. In traditional concepts of power system security, researches were conducted in restoring the power network and maintaining the operation of the system against electrical contingencies. However, with time and increasing complexity in power network, the complex power network has become susceptible to unplanned outages due to cyber-attacks and physical failures of key elements of the power system that includes grid system.

Preface

Large-scale disruptions of the power grid have occurred several times across the globe confirming the existence of inherent vulnerabilities. The grid network of Switzerland suffered a major setback due to electrical failures on 28th September 2003. North America also faced a major blackout in 2003. Such failures affect millions of people and cause losses of billions of dollars. Similar incidents were reported in Canada in 1998, in Europe in 2006, in United States due to Katrina, in 2007, etc. On 30th and 31st July 2012, the failure of the Indian grid affected 680 million people and loss of billions of dollars.

To minimise the economic, social and political impacts of power system outages, the grid must be sufficiently strong to face the vulnerability. The response of the grid against such diverse disruptions is a matter of concern. Indeed, it is clear that after decades of human and technological growth, problems are inevitable; however, it is important is respond to the problems. These challenges motivate research on the aspects of vulnerability analysis of power system networks.

To assist in developing tools or metrics that could be used for early identification of the zones of possible unplanned failure or attack, this book attempts to develop methodologies to identify the vulnerable elements, vulnerability of the network and its resiliency while encountering natural calamities or attacks at its key elements.

The modelling of power system as a complex network offers a framework to analyse the response of the grid system and distribution system while the network is subjected to the threat of failure and even the threat of cascading collapse. To achieve this target, different mathematical and computational techniques in line with the techniques used in complex network theory and some aspects of the concepts of probability have been utilised in this book.

Chapter 1 introduces the vulnerability analysis of power grids from a complex network perspective. In this chapter, the background of the work is described as the missing link between previous studies and how these details make the study of vulnerability analysis of power grids much more useful. A brief history of complex networks under this research domain are also presented. Moreover, how power grid is characterised with complex features is discussed in the chapter.

In Chapter 2, the concepts of bus admittance matrix and bus impedance matrix are discussed in detail for assessing the vulnerability of the transmission system employing complex network theory. In this chapter, traditional fault analysis approach in power system using symmetrical components has been utilised. Formation of Y_{bus} matrix as well as Z_{bus} matrix are explained in the chapter followed by determination of symmetrical fault current using $[Z_{bus}]$ inversion. Three phase balanced fault, Line to ground (L-G) fault, Line to line (L-L) fault, Double line to ground (L-L-G) and Open Conductor faults are analysed using $[Z_{bus}]$. Line currents during fault condition and open conductor faults are also determine.

In Chapter 3, the concept of contingency plays an important role in planning an islanding condition during emergency. Post-switching condition (transient state) of any component in power transmission system, any overload or overvoltage conditions are checked to predict the steady bus voltage and line currents. Hence, contingency analysis is necessary to assess every step after addition or removal of lines along with the use of Z_{bus} building algorithm. This chapter depicts cases of single as well as multiple contingencies. The concept of equivalencing of more than one network area by using tie lines and, consequently, inter-tie contingency analysis is described.

In Chapter 4, a comprehensive description of the parameters in complex network theory is presented employing a pure topological approach. The concept of incorporating specific electrical features of power systems is introduced in modelling of a power network as a complex network. Applying the mathematical concepts of complex network theory, an extended topological methodology (betweenness, vulnerability using global efficiency and netability) is discussed, which involves complex network theory and electrical features of power systems. Criticality assessment employing electrical centrality is also discussed, and subsequently, concepts of vulnerability indices are introduced.

In Chapter 5, vulnerability and criticality of power networks are assessed in grid-connected power system. Commonly used power system operating parameters have been utilised to rank critical lines based on the magnitudes of electrical betweenness. In the simulation, physical parameters of the power system that includes line flow, power limits and transmission capabilities are incorporated to analyse the vulnerability of the power network after obtaining the concept of netability using complex network theory. A new index called Grid Vulnerability Index (GVI) has been introduced in the context of advancing system performance.

Chapter 6 details the analysis of cascading failure and islanding in grid networks. In this chapter, power systems are modelled for unplanned outages and attacks leading to cascading failure. The electrical betweenness of transmission lines and/or nodes has been utilised to ascertain the criticality of transmission line(s) that might pose a potential threat of outage.

Considering the concept of preferential load distribution among adjacent lines, the simulation examines whether any line exceeds its ultimate capacity and becomes an overloaded line prone to cascading failure, subsequent to dynamic redistribution of load. The evolution of probabilistic tools along with their correlation with complex parameters illustrates their strong influence in modelling cascading failures. Cascading failure model is analysed using preferential probability ultimately leading to system islanding.

In Chapter 7, the immunity of a power network against unplanned or designated vulnerable attacks is defined. The concepts of complex network theory, integrated with statistical methods, are employed to determine the criticality of damage to the nodes and edges of topographical model of an electrical power system encountering a vulnerable attack on the power system. The capability of the network to resist fragmentation and mitigate and overcome stresses and failures are characterised by the resiliency of the network. Betweenness centrality of an electrical power network being one of the prime parameters of assessment for vulnerability of that network, computations have been conducted to obtain the moments of degree distribution of the network. This chapter highlights the quantification of critical fraction of damaged nodes as well as the percolation threshold of the power network and the characterisation of resiliency of a typical electrical power system following a routine electrical failure, outage or designated attack on that electrical power system. This chapter assesses metrics that resist and mitigate stresses and failures of the power system to ensure network survival in case of extreme events and natural or man-made disasters. Probabilistic approach has been applied to determine the loss of connectivity of the networks, and hence assessment of resiliency metrics in both transmission and distribution networks.

In Chapter 8, distributed generation is defined and described in detail including the most widely used renewable energy sources. Solar, wind, small hydro and tidal energy sources are discussed. The chapter explains how distributed renewable energy sources that are a part of the power system are a better alternative. Distributed generation has been applied to develop an effective mitigation strategy to enhance the robustness of the distribution system. In this process, simulations are conducted to obtain an optimal size and placement position for the distributed generation in the complex power network using distributed energy resources. The impact of distributed renewable energy sources on the *power loss* and *voltage* profile of sub-transmission network at different penetration levels are worth investigating in this chapter. This chapter shows how application of distributed renewable energy sources can reduce the criticality of the entire network. The benefit of installing distributed generation at the critical nodes (buses) of the same electrical power network has also been discussed.

The authors have attempted to present the topics covered in this book in a lucid language. Any constructive criticism and/or suggestion to further enhance the knowledge related to the topics presented will be highly appreciated.

Tamalika Chowdhury
Abhijit Chakrabarti
Chandan Kumar Chanda

Kolkata, India

Acknowledgments

The authors acknowledge the motivation and support given by the management of IIEST Shibpur and Jain College of Engineering, Belgaum, Karnataka, India. The authors also express their sincere thanks to Dr. Debraj Sarkar for his motivation and support in preparing the book. Dr. Tamalika Chowdhury further acknowledges the blessings and love of her parents Prof. (Rtd.) Tapan Kumar Chowdhury and Prof. Alpana Chowdhury. The authors finally mention the criticism and queries raised by their students and scholars while they teach the courses related to this topic. These criticisms and queries helped the authors to re-orient the mode of presentation of the topics covered in this book.

Tamalika Chowdhury
Abhijit Chakrabarti
Chandan Kumar Chanda

Authors

Tamalika Chowdhury Ph.D. (Engg.) is an assistant professor in Jain College of Engineering, Belagavi. She has been a research scholar in the Department of Electrical Engineering, Indian Institute of Engineering Science & Technology (IIEST), Shibpur (formerly BE College Shibpur) Howrah, West Bengal, India. She pursued her Ph.D. program under UGC-BSR Fellowship. She received her B.Tech and M.Tech degree in electrical engineering from West Bengal University of Technology (WBUT). Her research interests include power grid, power system vulnerability, manifestation of renewable energy-based distributed generation in power system and assessment of power grid resilience using complex network theory.

Abhijit Chakrabarti Ph.D. (Tech) is a professor of electrical engineering in IIEST, Shibpur. He is the former Head of the Department, Electrical Engineering, IIEST Shibpur and former Vice-Chancellor of Jadavpur University. He is also the former Vice Chairman of West Bengal State Council of Higher Education, Government of West Bengal. He has published 132 research papers in peer-reviewed international and national journals as well as in conferences. He has authored 17 books in electrical engineering. He is a member of EEAC of National Board of Accreditation, Government of India, as well as a visiting member of IET, UK in accreditation process of ADAMS, UK. He has visited foreign countries like Japan, United States and UK. He is a fellow of the Institution of Engineers, India and member of IET, UK. Professor Chakrabarti has received many awards and medals for his research in the field of power system engineering. He is a member of different boards and committees in universities in India and an expert member in numerous technical and standing committees. He has an active interest in research related to power system engineering, electrical machines, power electronics and electric circuits.

Chandan Kumar Chanda is working as a professor in the Department of Electrical Engineering, IIEST, Shibpur, India. He earned his Ph.D. degree from the Department of Electrical Engineering, B.E. College (DU), Shibpur, India with specialisation in power systems. Dr. Chandan Kumar Chanda has over 31 years of teaching and research experience in the diverse field of power systems engineering and almost 5 years of experience in industry. His areas of interest include smart grid, resiliency, stability and renewable energy. He is a recipient of Tata Rao Gold Medal. He is actively involved in various research projects funded by centrally funded organisations like DST and UGC. He has published 150 research articles in reputed national/international journals and conferences including 45 research papers in SCI/SCOPUS-indexed journals. He is a member of the editorial board and a guest editor of numerous reputed Journals. He has authored and co-authored four books with reputed publishing houses like Mc Graw Hill, PHI, etc. He has contributed five book chapters in *International Proceedings*. Ten research scholars have received their Ph.D.

degree under the supervision of Dr. Chanda. He has visited several countries including United States, UK, Australia, Japan and China for academic purposes. He is a senior member of IEEE (United States), member of IET (UK), Fellow of Institution of Engineers (I), C-Engineering (I) and life-member of ISTE.

1 Introduction

1.1 BACKGROUND AND MOTIVATION

Electrical energy has become one of the primary sources of energy. Consumption of electrical energy is increasing day by day. To meet this power demand the power-generating stations are growing in both size and capacity. However, smooth operation and uninterrupted power supply has become a big challenge for engineers, and with the growth of network size and capacity, the problem has increased in both severity and complexity.

The last decade has been very challenging for the energy sector, especially the electrical power system. It is not that the previous years' challenges were less but the consequences are much severe in today's world. The reason is very simple: the electrical system over decades has evolved in such a way that it has become a need for survival of the modern society as well as economic development. Integration of extensive and modern development in information and communications technologies has revolutionised the power system in terms of observability over the operation of the widespread power networks, and hence, the response has also become faster and efficient. However, this has created another challenge of cyber security to the power system. Natural disasters such as hurricanes, ice storms and earthquakes are continuously challenging power system researchers to find ways to reduce the damage caused by these events.

In the modern world, the civilisation depends upon bulk-scale usage of electricity, for which the infrastructure involves proper operation of the electric grid. The electric grid, commonly known as power grid, is also equipped with advanced and intelligent computer-based sensing and automation system as well as information technology-related software. This network possesses an efficient communication system and distributed energy resources in addition to conventional equipment and transmission lines. Integrating these new technologies has resulted in more interconnections and interdependencies between the physical and cyber components of the grid. Natural calamities and electrical contingencies have been the traditional forces of disruption of part or whole of the grid. However, complexities in the political world and threats on society from terrorist outfits have given rise to the apprehension that there is a risk to electric grid from intentional attacks. The vulnerability of infrastructure networks to electric grid outages is becoming a major global concern.

Power systems, composed of grids and sub-grids, support the generation, transmission and distribution of electricity. Power grids can be vulnerable to attacks and failures. There are instances where failure of component(s) in the grid or line(s) have caused partial or complete grid collapse. It is the need of the day that the utilities must maintain high quality of customer service. Therefore, it is essential to minimise such problems and fix them quickly if they occur. Therefore, intensive research is needed to study the robustness and vulnerability of a power network. *Complex network theory* has been investigated extensively in recent years due to its potential for solving large-scale practical problems. In modelling of different complex systems such as biological systems, social systems can be modelled by the use of complex network theory. While power system researchers are working towards the upgradation and safety of the grid and sub-grid, the research on power grid and sub-grid is an interdisciplinary problem, crossing over several disciplines. The structure of the network of transmission lines and buses stores a huge amount of useful information that can be used for the benefit of the society through the science of complex networks. It is with this goal that researchers have modelled the power grid and sub-grid as a complex network to analyse the dynamics of the grid and sub-grid from a network's perspective, suggesting some strategies that can help strengthen the grid and sub-grid.

The developments in the area of application of complex network theory in real power systems have facilitated advanced research in power system. Following the concepts described in complex network theory, a power system can be modelled as a graph with nodes and vertices and further analysis can help in identifying the critical lines and locating faults. It is an attempt to add knowledge to the existing techniques for structural vulnerability analysis of power networks.

The modelling of complex networks has significantly advanced in the last decade and has provided valuable insights into the properties of real-world systems by evaluating their structure and construction. Several phenomena occurring in real technological and social systems can be studied, evaluated, quantified and remedied using network science. The electric power grid is one such real technological system that can be studied through the science of complex networks. The electric grid consists of three basic subsystems: generation, transmission and distribution. Power transmission lines and feeders are the key components of transmission and distribution systems. The power grid infrastructure transfers electricity from the generation to the load end through a complex configuration of transmission lines, buses and different power equipment. Some of these elements of the network are more important than others either due to their location in the network, due to the load they carry or due to their inherent characteristics. If these key elements fail due to outage or attack, then the performance of the entire system can be affected significantly. This book is motivated to identify such critical and vulnerable elements in the power system using some concepts from the complex network theory. Attempts have been made to highlight mitigation strategies so that the resiliency of the network is improved, the power network can successfully encounter unplanned outages and attacks and the system can survive from cascading collapse.

1.2 GROWTH OF MODERN LARGE-SCALE POWER SYSTEMS

Modern power systems are large-scale power grids belonging to a typical class of complex systems, whose characteristics can be summarised as follows:

i) **Power supply and consumption should be balanced instantaneously**: Electricity is to be generated and transmitted by lines, transformers and switches to end-users, with the speed of light. So far, it is yet impossible to achieve industrial-scale and high-capacity storage of electrical energy.

Hence, one critical feature of power systems is that the production (generation) and consumption are completed almost simultaneously, and thus, the power supply and demand need to be balanced instantaneously. If the balance is broken, the system will lose its stability, which will lead to power outages of different scales.

ii) **Large-scale power**: A modern power system consists of three subsystems: the first system, where the energy conversion, transmission, distribution and consumption take place. The second one is the automatic control system, also called the secondary system, which is responsible for the secure, stable and economic operation of the power system. And the third is the power trading system. The secondary system usually has more components than the primary system as, in addition to the communication equipment, a large number of sensors are installed to monitor various states, such as voltage, current and temperature, of the devices in the primary system. Since the last couple of decades, the interconnection of regional power systems has attempted to optimise the use of hydroelectric, thermal and nuclear powers together with other energy resources inviting more constraints into the picture.

iii) **Modern power system components are complex**: The power system components are not only *huge* and of numerous kinds but the components have different complexities as well. Generators and motors that are based on electromagnetic induction law and Newton's second law are the most common components that are used to realise the conversion between electrical and mechanical energies. Others components such as boilers, pressure pipes, coal mills and reactors help to convert chemical, hydraulic and nuclear energies to mechanical energy. Post 1990s, with the rapid development in electronic technologies, high-voltage direct current transmission (HVDC), thyristor-controlled series compensation and static VAR compensator as well as other FACTS devices have been utilised widely, which makes the electric components in power systems more complicated. Different from traditional power equipment, their main functionality is to flexibly switch the power electronic devices to control power transmission. Loads are the most complicated class of components in power systems. The modelling of dynamic loads has not been fully solved yet. Researchers usually classify loads into constant impedance loads, temperature-controlled loads, compressor loads and other simplified models. In addition, various protections and controls are important classes of complex components in power systems. For example, for a generator, there are usually excitation controllers, speed controllers, low-frequency protectors, under- and over-excitation limiters. There are also centralised control systems, such as Automatic Generation Control and Automatic Voltage Control, located in dispatch centres to coordinate and control the overall performance of the entire system.

iv) **Factors affecting secure operation of power grid:** Climatic changes, human activities and natural calamities such as earthquakes, lightning, snow and storms force power systems to be constantly exposed to various random perturbations. For example, lightning or untrimmed trees may cause short-circuit faults on transmission lines, and may destroy electricity poles and substations. Due to hot weather, more people use air conditioners; and major sport and political events may change the type of power loads. Such random factors increase the complexity in operating and managing power systems. In addition, as electricity markets are deregulated, energy management systems become more complicated. In combination with the random effects just described, the planning, operation and management of power grids are facing demanding challenges.

From the above four aspects, it is clear that modern power systems are large-scale complex systems. Therefore, naturally, they become one subject in complexity studies. Although modern power systems have complex dynamic characteristics, their robustness and reliability are relatively high due to the implementation of various protections and controls. However, serious losses were caused in the few power failures reported in the past years. The interconnection between different areas/zones brings both great social and economic benefits as well as risk for failures, especially cascading failures and blackouts. Large-scale disruptions of the power grid have occurred several times across the globe confirming the existence of inherent vulnerabilities. The grid network of Switzerland suffered a major setback due to electrical failures on 28th September 2003. North America also faced a major blackout in 2003. Such failures affect millions of people and cause losses of billions of dollars. Similar incidents have occurred in Canada in 1998, in Europe in 2006, in United States due to Katrina, in 2007, etc. On 30th and 31st July 2012, the failure of Indian grid affected 680 million people and loss of billions of dollars.

To minimise the economic, social and political impacts of power system outages, the grid must be sufficiently strong to face the vulnerability. The response of the grid against such diverse disruptions is a matter of concern. Indeed, one thing is clear that after decades of human and technological growth, problems are inevitable; however, it is important to respond to these problems. These challenges motivate to work on the aspects of vulnerability analysis of power system networks.

Modern power system security is prone to unplanned outages due to natural calamities and cyber threats that may lead to destructive consequences for the power system network. In traditional concepts of power system security, researches restored the power network and maintained the operation of the system against electrical contingencies. However, with time and increasing complexity in the power network, the complex power network is much more susceptible to unplanned outages as well as to cyber-attacks and physical failures of key elements of the power system that includes both grid and distribution systems.

To assist in developing tools or metrics that could be used for early identification of the zones of possible unplanned failure or attack, the upcoming chapters state methodologies to analyse the power system faults and contingencies in a classical way and to identify the vulnerable elements, vulnerability of the network and its resiliency while encountering outages due to natural calamities or attacks at its key elements.

1.3   COMPLEXITY IN POWER SYSTEMS

In recent years, the difference between complexity and complication has been noticed by power system researchers. A few studies regarding complexity and applications of its theories to power systems have been reported. Instead of looking at the details of particular blackouts, some researchers [1] studied the statistics and dynamics of a series of blackouts with approximate global models. However, some researchers [2] employed topology analysis to figure out the vulnerability of a given transmission system and concluded that when a network is attacked following a delicate sequence corresponding to their criticality, the network would illustrate more vulnerability.

Energy infrastructures, such as power systems, are characterised by a large number of components and many different types of interactions among them. Size itself does not infer complexity. Continental-scale power grid, for example, is the biggest dynamic system in the world, but from a physical viewpoint, it can be modelled by a huge set of differential and algebraic equations. It may conjure complexity in the computational efficiency; however, it is somehow solvable using computationally powerful facilities and advanced algorithms. In contrast, complexity arises when the physical substrate interacts with the rest of hierarchical levels governing and using the infrastructure. The overall expected performance and dynamic evolution are related to interactions at the "individual" scale. These phenomena can neither be handled nor studied with a set of equations in any form. Studies and applications related to the deregulation towards market environment have thrived in both academia and industries. This change brought a great challenge to power systems in production and transmission. With the prevalence of distributed generation and smart grids, the distribution and utilisation are confronting new scenarios in which a large number of users have transformed from passive recipients to active participants. The emerging situation and newly introduced players with clear self-interest display an important role for the future power system, which will further increase the complexity of the power system.

The complexity of power systems also increases with the change in its administrative mechanism. Initially, in power systems, each utility and/or pool of utilities has control centres which support today's hierarchical monitoring and grid control. Moreover, electrical market is gradually introduced into power systems to transmit the least expensive power in power grids. An important consequence of this situation is that utilities require systematic integration of monitoring, computing and controlling for improved performance. Therefore, the interaction between power grids and decision information via a cyber layer is more complex than before.

Furthermore, renewable energy, such as wind power, solar energy and fuel cell, is drastically emerging and developing in the distribution level of traditional power industries. This trend increases the complexity of the overall power systems.

1.3.1   COMPLEXITY AND COMPLEX SYSTEMS

There is no general accepted formal definition of complexity science as Heyligen stated: "Conceptually, the most difficult aspect of complexity is still its definition and the deeper understanding that goes with it" [3]. There have been many different endeavours in complexity over almost all disciplines, covering various definitions and measures of complexity. In general, they can be categorised into three groups: Algorithmic complexity [4,5], deterministic complexity [5] and aggravated complexity [5]. The first group covers the complexity of describing system characteristics, such as mathematical complexity theory and information theory. The second group includes the interaction of very few key variables that create largely stable systems prone to sudden discontinuities, such as chaos theory and catastrophe theory. The last one mostly concerns how individual elements work in a synergy that generates complexity in a system.

To have a lucid understanding of complexity, as well as generalising the concept without losing any useful positive meaning, complexity can be defined as follows:

"Complexity is a property that makes it difficult to analytically formulate its overall behaviour even when knowing the complete information about its elements and their relationships." Here "difficult" can involve several aspects such as size, depth, computational indication, efforts in a search for the most apt representation, etc.

Accordingly, a general and logical definition of complex system is one that exhibits complexity: "A system, that can be decomposed in a set of different types of elementary parts with autonomous behaviours, goals and attitudes and an environment, is complex if its modelling and related simulation tools cannot be done resorting to a set of whichever type of equations expressing the overall performance of the system, in terms of quantitative metrics, or a function on the basis of state variables and other quantitative inputs" [23]. This definition is an articulated and practical way targeting at engineering systems.

Over the last several years, complexity science has changed the way scientists approach all fields of life form biology to medicine, and from economics to engineering [6–10]. The concepts or techniques such as self-organisation, genetic algorithm, cellular automata, criticality, artificial life or chaos theory are now widely accepted and used as new means of understanding the constantly changing reality. The history of complex systems research including these concepts begins in the 1950s, emerging with the advent of von Bertalanffy's systems theory, the appearance of nonlinear phenomena in scientific fields away from physics, such as chemistry and biology, and the study of feedback concepts in communication and control in living organisms, machines and organisations. From these early stages, the idea of threshold became the cornerstone of much of the complexity science developments of the 1980s, especially in the cellular automata and artificial life fields, where complex behaviour seemed to appear suddenly [11–13]. Since then, many books, journals, conferences and even whole institutes devoted to the field have flourished everywhere, and even computer modelling of complex systems has become widely accepted as a valid scientific activity. Marti Rosas-Casals [32] cited a conceptual framework in which complexity pervades both the (i) structure (i.e. formal arrangement of the constituent parts), (ii) dynamics (i.e. functional behaviours) and (iii) evolution (i.e. the way it has reached its actual formal and functional state) of any system. It covers the major aspects of complexity science and different technologies and methods. These methods and techniques of complex systems are grouped into three categories: (i) those for analysing data, (ii) those for building and understanding models and (iii) those for measuring complexity.

With new advances in complex network theory, most of the complex systems in the world can be modelled and described in the form of a complex network. Several network models have been proposed with the intention of studying the topological characteristics and behaviour of such complex systems. The types of networks can be broadly classified as *regular, small-world, random* and *scale-free networks.*

1.3.1.1 Regular Networks

Initially, complex networks were assumed to be completely regular. A few examples of regular networks are chains, grids, lattices, fully connected graphs, etc. The relatively simple architecture allows us to focus on the complexity caused by the nonlinear dynamics of nodes and edges without the additional complexity of the topology itself [14].

Hence, these network models have been used quite often to study dynamic systems such as disease spread and ecosystems. For a regular network, the clustering is high and the average distance between nodes is high [15].

1.3.1.2 Random Networks

Regular network models were not efficient to describe the phenomenon of real-world systems; hence, Erdos and Renyi [16] came up with the idea of random networks. The term random refers to disordered arrangement of links connecting various nodes. Such networks are created when each pair of nodes is connected by a link with uniform probability. These networks were studied by Erdos and Renyi from a purely mathematical viewpoint. However, networks with complex topology and unknown properties are often represented as a random graph. Hence, the random graph theory has a significant place in the study of complex networks. In case of random networks, the average distance between nodes is small and the clustering is low [15]. Such networks are usually robust to targeted attacks but very vulnerable to random failures or attacks [17].

1.3.1.3 Small-World Networks

There are networks whose behaviour falls between a regular network and a random network. These types of complex networks are classified as small-world networks, which were first introduced by Watts and Strogatz [15,18,19]. Even though there are few long-distance connections present in such networks, the shortest path length between any two nodes scales logarithmically or at a smaller rate with increasing network size. This means even in a small-world network with many nodes, the shortest path length between two nodes is likely to be relatively small with high clustering [20]. Power system is a good example of a small-world network.

1.3.1.4 Scale-Free Networks

About the same time when Watts and Strogatz developed the small-world model, Barabasi and Albert [21] came up with an alternate network model which grew by *preferential attachment* and were called scale-free networks. These networks grow in such a way that nodes with higher degrees receive more new connections compared to others with low degrees, that is, the probability of making a new connection to a node is proportional to its degree. Unlike the Gaussian distribution which has cut-off values where the distribution approaches zero, scale-free distributions have no such cut-offs, and instances of all scales are present; hence, the term scale-free. These networks are simply those with *power law degree distribution* where most nodes have low connectivity but some are highly connected to the rest of the network [21]. This makes those highly connected nodes or hubs very vulnerable to attacks. Thus, such network models are robust to random attacks but can be highly vulnerable to targeted attacks [17]. A classic example of a scale-free network is the Internet.

Figure 1.1 illustrates simple examples of regular, random, small-world and scale-free networks [22].

1.3.2   POWER GRID AS A COMPLEX SYSTEM

Power grids have been widely acknowledged as a typical complex network because of both their huge sizes of components and the complex interactions among them. Complex network theory has received considerable attention recently which has been used in many different fields. Numerous studies [23] including basic characteristics, statistical global graph properties, small-world property, scale-free property, degree distribution, betweenness distribution and vulnerability analysis have been performed for power grids as they are infrastructures in our society. There is a strong link between the topological structure and operation performance in power systems because the structural change could alter operational condition of a power system as well as its operation performance.

Consequently, there is an increasing interest in analysing structural vulnerability of power grids through complex network methodology. With the development of complex system theory, power grids arise as natural objects of investigation under the conceptual framework of complex systems, particularly as complex networks. Therefore, complex network methodology as one of approaches to study complex systems has been used to analyse and understand power systems from a topological viewpoint. How complex network theory and methodology are applied in power systems study, especially in vulnerability analysis, will be addressed in the subsequent chapters.

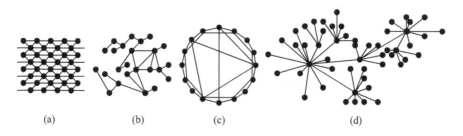

(a)   (b)   (c)   (d)

FIGURE 1.1   Different types of complex networks: (a) Regular, (b) random, (c) small word, and (d) scale free.

1.4   TRANSMISSION SYSTEM FAULTS

In an electrical power system, *a* **fault** *is any abnormal flow of unavoidable and undesirable current which can temporarily disturb the stable operating condition of the system due to insulation failure by any internal or external agent at any point of time.* The agents causing fault may be natural (lightning, wind damage, trees falling across transmission lines, rodents, squirrels or birds shoring lines) or man-made (outage due to islanding, vehicles or aircraft colliding accidentally with transmission towers and poles, any kind of intentional attacks or vandalism). The nature of the faults may be any abnormal condition which reduces basic insulation strength between phase conductors, or between phase conductors and earth, or between any earthed screens surrounding the conductors [24]. In practice, a reduction is not regarded as a fault until it results either in an excess current or in a reduction of the impedance between conductors, or between conductors and earth, to a value below that of the lowest load impedance normal to the circuit.

The faults in transmission system causes overcurrent, undervoltage, unbalance of phase, reversed power and high-voltage surges. This results in failure of equipment, electrical fires and interruption in normal operation of the network. These faults occur due to the failure of one or more conductors. The most common faults that occur in power system are unsymmetrical or asymmetric which lead to unequal currents with unequal phase shifts in a three-phase system. This kind of fault occurs in a system due to the presence of an open circuit or short circuit of transmission or distribution line. The path of the fault current may have either zero impedance (short circuit) or nonzero impedance [25]. Other type of fault includes one conductor or two conductors open (open conductor faults). Such faults occur when conductors break or when one or two phases of a circuit breaker is inadvertently open.

There are mainly three types of faults, namely, line-to-ground (L-G), line-to-line (L-L) and double line-to-ground (LL-G) faults. Line-to-ground fault (L-G) is the most common fault, and 65%–70% of faults are of this type. It causes the conductor to make contact with earth or ground. Line-to-line is a short circuit between lines caused by ionisation of air, or when lines come into physical contact, for example, due to a broken insulator. In transmission line faults, roughly 5%–10% are asymmetric line-to-line faults. Line-to-ground is a short circuit between one line and ground, very often caused by physical contact, for example, due to lightning or storm damage.

1.5   CONVENTIONAL CONTINGENCIES IN POWER TRANSMISSION NETWORK

Currently, due to large interconnection and stressed operation, power utilities are facing severe problems of maintaining the required security. Today more emphasis is paid to the greater utility of generation and transmission capacity, which has caused the system to operate much closer to their limits. So, it has become indispensable to do voltage security assessment accurately and instantaneously to prevent the system from voltage collapse. The concept of security in system operation may be divided into three components: monitoring, assessment and control. Security monitoring starts with measurement of real-time system data to provide up-to-date information of the current condition of power system. *Security assessment* is the process whereby any violation of the actual system operating is determined. The conventional methods for security assessment are based on load flow solution where full AC load flow is made to run for all contingencies. Load flow constitutes the most important study in a power system for planning, operation and expansion. The purpose of load flow study is to compute operating conditions of the power system under steady state. These operating conditions are normally voltage magnitudes and phase angles at different buses, line flows (MW and MVAR), real and reactive power supplied by the generators and power loss. The second and much more demanding function of security assessment is contingency analysis. Operations personnel must know which line or generation outages will cause flows or voltages to fall outside limits. To predict the effects of outages, contingency analysis techniques are used.

Contingencies are defined as potentially harmful disturbances that occur during the steady state operation of a power system. The contingencies are in the form of network outage such as line or transformer outage or in the form of equipment outage. The outage considered in transmission network are line outages. Outages that are important from a limit violation viewpoint are branch flow for line security or MW security and bus voltage magnitude for voltage security. The conventional methods for security assessment are based on load flow solution where full AC load flow is made to run for all contingencies. *Contingency procedure* includes failure events one after another in sequence until "all credible outages" have been studied. For each outage tested, the contingency analysis procedure checks all lines and voltages in the network against their respective limits.

1.6   THREATS AND THEIR CONSEQUENCE IN POWER SYSTEM OPERATION

The science of power grid technology is progressing worldwide. Most are investing to transform their traditional power grid to smart grid. They have started realigning their organisation to support a *smart grid* vision. At its core, a smart grid utilises digital communications and control systems to monitor and control power flows, with the goal of making the power grid more resilient, efficient and cost-effective. *Smart grids* increase the connectivity, automation and coordination between suppliers, consumers and network by modernising grid features like demand side management, generation, real-time pricing and automated metre activation and reading. As smart grid technology is advancing with time, complexity in managing the power grid is also increasing.

The grid must be sufficiently strong to face the vulnerability. The response of the grid against such diverse disruptions is a matter of concern. Indeed, one thing is very clear that after decades of human and technological growth, problems are inevitable, but it is important to respond to the problems. These challenges motivate to work on the aspects of vulnerability analysis of power system networks.

Modern power system security is prone to unplanned outages due to natural calamity and cyber threats that may lead to destructive consequences for the power system network. In traditional concepts of power system security, researches were conducted in restoring the power network and maintaining the operation of the system against electrical contingencies. However, with time and increasing complexity in power network, the complex power network has become susceptible to unplanned outages as well as cyber-attacks and physical failures of key elements of the power system that includes grid system and the distribution system.

A cyber attacker is an intelligent agent capable of coordinating attacks that result in deliberate component failures, which increases the risk of low-probability, high-impact contingency scenarios. The grid network or the distribution network is also vulnerable while encountering unplanned contingencies during natural calamity. The traditional physical contingency analysis tools rely on a version of the power flow problem, which requires calculating the solution of operating parameters for all buses in the power system. While personnel within a utility may have knowledge of specific bus injections and voltage magnitudes needed to solve the power flow problem, it is not uncommon that a cyber attacker plans to identify critical elements in the power network and fulfils their evil desire in making the power system vulnerable to attack at its critical elements. In addition, the power system faces high degree of vulnerability when the system is subjected to unplanned electrical failures of a few of its key elements during natural calamities.

Instances like faults at power stations, damage to electric transmission lines, substations or other parts of the distribution system, a short circuit, cascading failure, fuse or circuit breaker operation may cause complete collapse of the power grid. In addition to these, there has been instances of several targeted attacks on grids in power system networks in an attempt to collapse the entire system network [33]. To maintain the reliability of power systems, wide area monitoring systems are exploited to obtain real-time system status, which is essential for the maintenance and control of power systems. Most studies on security issues in smart grid have focused on how to protect data transmission in power grid. Accordingly, the attack on the measurements is called *False-data Injection Attack* (FIA) [26]. FIA allows attackers to manipulate the estimated state of the power grid by changing the measurements on a fraction of monitoring devices. When the attacker accesses the power grid communication network, the attack can bypass the existing bad measurement detection.

The vulnerability assessment of power transmission networks employing complexity science investigates the critical targeted vulnerable areas of such an attack. Hence, it is a present-day requirement to be equipped with proper knowledge on vulnerability of grid networks.

1.7 MODELLING OF POWER TRANSMISSION NETWORK AS COMPLEX NETWORK

A network can be defined as a set of nodes or vertices with connections called links or edges [27]. A *vertex* is the fundamental unit of a network, also called a *node* (in electrical and computer science). They are connected together by lines called *edges*, also known as a *link* (in electrical and computer science). The nodes or vertex represent various elements, tangible or otherwise, such as hardware devices, buses, generators and transformers, and the edges represent the relationship between these elements or the way they interact with each other. Figure 1.2 shows a network with nodes and edges in its simplest form.

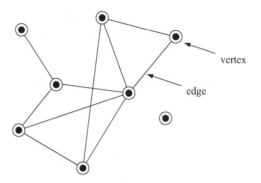

FIGURE 1.2   A network with nodes and edges.

Networks can be of different types: For instance, there can be networks with similar nodes and edges or there can be networks with more than one type of nodes and different types of edges. Further, these vertices and edges can have various properties. For example, edges can have weights associated with them which might represent how strongly or loosely any two nodes are connected. Such networks are called *weighted networks*. Any information is transferred within the network via the nodes using connecting links. Sometimes, the flow of this information can only be in one direction, in which case the network is termed as *directed graphs* or *digraphs*. These directed networks can either be *cyclic* containing a closed loop of edges or *acyclic* containing no such loops. Then, there are *undirected networks* in which flow of information can be in both directions of connections. Figure 1.3 shows

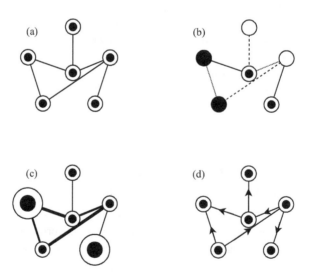

FIGURE 1.3   Examples of different types of networks. (a) Network with Identical nodes and edges, (b) Network with different types of nodes and edges, (c) Network with different weights put on to nodes and edges, (d) An example of Directed graph.

simple examples of various types of networks. Figure 1.3a shows a network with identical nodes and edges, Figure 1.3b shows different types of vertices and connections, Figure 1.3c represents a network in which nodes and edges have different weights associated with them and Figure 1.3d shows an example of a directed network.

To represent the power grid as a complex network, an unweighted and undirected graph composed of nodes and edges needs to be built first. Using metrics from graph theory and modern complex networks analysis, the results provide insight into the properties of power grids, considering only topological information. In summary, the goal of this model is to characterise the topological structure of the *power transmission grid*.

To study power grids with this model, some simplifications are necessary. In the undirected graphs, each node represents a bus. It is important to note that in the physical grid, these buses can have different electrical properties; however, in this chapter, nodes are assumed to be homogeneous. This model ignores whether generators, loads, transformers or transmission lines connect to the bus. In the same manner, all transmission lines are modelled as edges with equal weight. Physical length and electrical impedance are ignored in the undirected graph representation.

The graph theory is the basic concept from which complex network theory has been derived. The power grid can be abstracted into the complex network with undirected graph $G = \{V, E\}$. It consists of two sets V and E, where the elements of $V = \{v_1, v_2, ..., v_N\}$ are the nodes (or vertices, or buses) of the graph G, while the elements $E = \{e_1, e_2, ..., e_L\}$ are its links (or edges or lines). The total number of nodes and links of the graph are N and L, respectively. The association of nodes with each other can be shown using adjacency matrix. If an edge e_{ij} exits between two nodes i and j, then the adjacency matrix A having order $N \times N$ *whose* entry a_{ij} becomes one and zero otherwise. *Adjacency matrix* is a $N \times N$ binary matrix in which the value of **[i, j]**th cell is **1** if there exists an edge originating from **i**th vertex and terminating to **j**th vertex, otherwise the value is **0**.

A power system can be considered a large complex network with nodes and edges. The generators, bus bars and loads can be identified as the nodes and the connecting transmission lines can be modelled as the edges or links. Let us consider a sample network consisting of five buses and seven lines modelled as an undirected graph.

Figure 1.4 shows the undirected graph of a system with five nodes and seven edges. It can be modelled using the principles defined above and mathematically represented as a matrix shown below:

$$A = \begin{bmatrix} 0 & 0 & 1 & 1 & 0 \\ 0 & 0 & 1 & 0 & 1 \\ 1 & 1 & 0 & 1 & 1 \\ 1 & 0 & 1 & 0 & 1 \\ 0 & 1 & 1 & 1 & 0 \end{bmatrix}$$

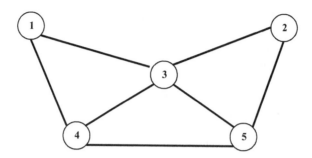

FIGURE 1.4   Undirected graph of a sample network.

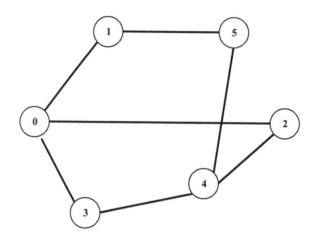

FIGURE 1.5   Undirected graph for Example 1.1.

Example 1.1

Problem Statement: Consider a graph given in Figure 1.5 below. Find the adjacency matrix for the network.

Solution

In this undirected graph, there are **N** vertices numbered from 0 to $N-1$ and **E** number of edges in the form **(i, j)**, where **(i, j)** represent an edge originating from i^{th} vertex and terminating on j^{th} vertex. Now, the adjacency matrix will be:

	0	1	2	3	4	5
0	0	1	1	1	0	0
1	1	0	0	0	1	1
2	1	0	0	0	1	0
3	1	0	0	0	1	0
4	0	1	1	1	0	1
5	0	1	0	0	1	0

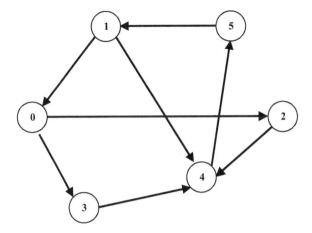

FIGURE 1.6 Directed graph for Example 1.2.

In the directed graph, an edge is represented by an ordered pair of vertices (i, j), in which edge originates from vertex i and terminates on vertex j.

Example 1.2

Problem Statement: Consider the same sample network in Example 1.1 but as directed graph, as shown in Figure 1.6. Find the adjacency matrix.

Solution

The adjacency matrix is given by:

	0	1	2	3	4	5
0	0	0	1	1	0	0
1	1	0	0	0	1	0
2	0	0	0	0	1	0
3	0	0	0	0	1	0
4	0	0	0	0	0	1
5	0	1	0	0	0	0

Furthermore, we can add weights to the connecting links which could be a measure of electrical or topological property depending on the application. In that case, the 1's will be replaced by the respective weights of the links. This will be further illustrated through various examples in this section.

WEIGHTED GRAPH

A graph is called a *weighted graph* when it has weighted edges, which means there are some "value" associated with each edge in the graph. For example, consider Figure 1.7 where there are weights associated with each edge.

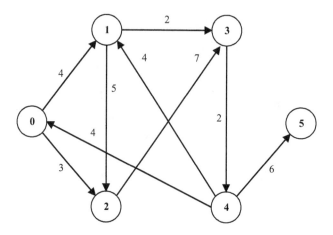

FIGURE 1.7   Directed weighted graph.

Each edge of a graph has an associated numerical value called a weight. Usually, the edge weights are non-negative integers. Weighted graphs may be either *directed* or *undirected*. The weight of an edge is often referred to as the "cost" of the edge.

Example 1.3

Consider an undirected weighted graph as shown in the Figure 1.8 and find its adjacency matrix.

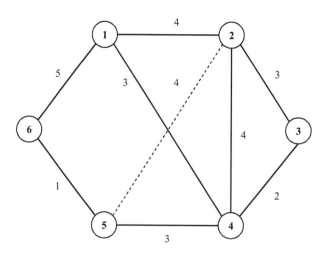

FIGURE 1.8   Undirected weighted graph of Example 1.3.

Solution

The adjacency matrix for this undirected weighted graph is given by

$$A = \begin{bmatrix} 0 & 4 & 0 & 3 & 0 & 5 \\ 4 & 0 & 3 & 4 & 4 & 0 \\ 0 & 3 & 0 & 2 & 0 & 0 \\ 3 & 4 & 2 & 0 & 3 & 0 \\ 0 & 4 & 0 & 3 & 0 & 1 \\ 5 & 0 & 0 & 0 & 1 & 0 \end{bmatrix}$$

Example 1.4

Consider a directed weighted graph as shown in the Figure 1.9 and find its adjacency matrix.

Solution

The adjacency matrix for this directed weighted graph will be:

$$A = \begin{bmatrix} 0 & 4 & 0 & 0 & 0 & 0 \\ 0 & 0 & 0 & 4 & 0 & 0 \\ 0 & 3 & 0 & 0 & 0 & 0 \\ 3 & 0 & 2 & 0 & 3 & 0 \\ 0 & 4 & 0 & 5 & 0 & 1 \\ 5 & 0 & 0 & 0 & 0 & 0 \end{bmatrix}$$

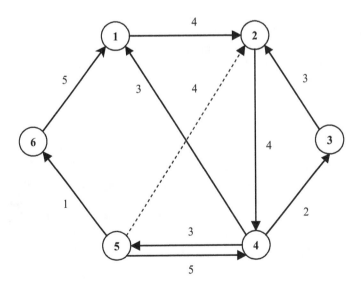

FIGURE 1.9 Directed weighted graph of Example 1.4.

1.8 STRUCTURAL PROPERTY ANALYSIS OF POWER TRANSMISSION NETWORK

As mentioned in Section 1.3.1, there are three main models of complex networks: small-world, scale-free and random networks in addition to regular networks.

To explain the characteristics of these different networks, Figure 1.10 shows a simple rewiring diagram [], which illustrates the relationship between regular, small-world and random networks. We start with a ring lattice with $n = 20$ nodes with each of them connected to four of their neighbours. Let each edge be rewired randomly with a probability P, that is, P is the ratio of number of lines rewired randomly versus total number of lines. Then, for $P = 0$, the original lattice is unchanged.

As the value of P is increased, the network becomes increasingly random, and for $P = 1$, all the lines are rewired randomly. The small-world phenomenon exists in the intermediate region $0 < P < 1$.

Different network will exhibit different structure and vulnerability property. Therefore, the first question to analyse power grid is what type of power grid is. The first reference as deduced by Watts and Strogatz [28] is that the western US power grid seemed to be a *small-world network*. Later, Barabasi and Albert [21,29] first published that degree distribution of a power grid was *scale-free* following a power law distribution function; however, few subsequent references supported this finding. Exponential cumulative degree function was detected in Californian power grid [30] and the entire US grid [31]. The topological features of the Union for the Co-ordination of Transport of Electricity (UCTE) power grid and its individual nation grids were analysed, and the results showed that these national transmission power grid topologies are similar in terms of mean degree and degree distribution, suggesting similar topological constraints, mostly associated with technological considerations and spatial limitations [32]. Furthermore, the topologies of the North American eastern and western electric grids were analysed to estimate their reliability based on the Barabasi–Albert network model. The results were compared to the values of power system reliability indices previously

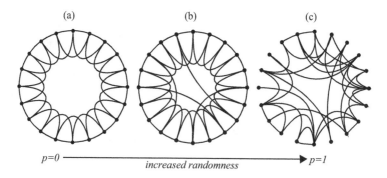

FIGURE 1.10   Relationship between network models and randomness. (a) Regular, (b) small word, and (c) random.

obtained from some standard power engineering methods, which suggested that scale-free network models are applicable for estimating aggregate electric grid reliability [33].

1.9 COMPLEX NETWORK APPROACH TO VULNERABILITY ASSESSMENT OF POWER TRANSMISSION NETWORK

The vulnerability analysis of network is the main motivation for the studies involving complex network analysis into power grids. When vertices are removed randomly and in decreasing order of their degrees for both generation vertices and transmission vertices, a connectivity loss is observed, which measures the decrease of the ability of distribution substations to receive power from the generators. The loss of generating substations does not significantly alter the overall connectivity of the grid owing to a high level of redundancy at the generating substations. However, the grid is sensitive to the loss of transmission nodes. Even the removal of a single transmission node can cause a slight connectivity loss. Especially, the connectivity loss is substantially higher when intentionally attacking higher-degree or high-load transmission hubs. The transmission highly connected hubs guarantee the connectivity of the power grid; however, they are also its largest liability in case of power breakdowns.

In European power grids, the topological properties of the Spanish, Italian and French power grids are studied and compared, finding those components whose removals seriously affected the structure of these graphs [34]. They also treated power grid as a simple graph with taking physical features into consideration; we think that the power grid vulnerability results obtained with this approach would be different from the real situation. Rosato et al. studied the topological properties of high-voltage power grid in Italy (380 kV), France (400 kV) and Spain (400 kV) [35]. An assessment of the vulnerability of the networks was done by analysing the level of damage caused by a controlled removal of links. Such topological studies could be useful for vulnerability assessment and for designing specific action to reduce topological weaknesses. Because the grids are the same as used in the former case, most of the results are consistent. Robustness of the entire European power grid was studied where the resilience against the failures and attacks of each national power grid was included. The authors [35,36] found that the European power grid composed of 33 EU power grids could broadly be classified into two separate groups: fragile and robust.

While analysing the first cascading failures in electrical power grid of the western United States [37], the degree distribution in this network appeared exponential and was relatively homogeneous. The distribution of loads, however, was more skewed than that displayed by semi-random networks with the same distribution of links. This implied, to a certain extent, that the power grid may have structures that are not being captured by the existing complex network models. Consequently, global cascades are probably triggered by load-based intentional attacks than by random or degree-based removal of vertices. The attack on a single vertex with large load may make the largest connected component reduce to less than one-half of its initial size, though the network is highly tolerant. It was observed [38] in North American power grid that the loss of a single substation can lead to a 25% loss of transmission efficiency caused by an overload cascade in the network. A systematic study of the damage caused by the loss of vertices suggested that 40% of the disrupted transmission substations may lead to cascading failures. While the loss of a single vertex can exacerbate primary substantial damage, the subsequent removals only make the situation worse.

Large-scale blackouts can be triggered by the failure of vertices with high loads. Perhaps it is because some highly connected vertices may not be necessarily involved in a high number of paths. A counterintuitive result was found that the attack on the vertices with the lowest loads is more harmful than the attack on the ones with the highest loads.

Within the framework of cascading failures in power grids using a dynamical flow model based on simple conservation and distribution laws, the role of the transient dynamics of the redistribution of loads towards the steady state after the failure of network edges was studies. It was found that considering only load flows in the steady state gives a best case estimate of the robustness; the worst case of robustness can be determined by the instantaneous dynamic overload failure model. The results of Norwegian high-voltage power grid showed that the size distribution of power blackouts in Norwegian power grid seems to follow a power law probability distribution.

1.10   CYBER SECURITY AND VULNERABILITY TOWARDS CASCADED COLLAPSE

Cyber security based vulnerability assessment measures tend to come in two forms: those that assess the vulnerability of cyber assets in the power system for compromise, and those that assess the consequences to the physical power system resulting from compromised cyber assets. However, because the electric grid is a cyber-physical system, the distinction between cyber vulnerabilities and the consequence to the physical system is often difficult to establish. For example, the existence of vulnerabilities in a cyber asset such as a control, protection or monitoring device or system does not necessitate a severe reliability impact to the physical power system if the asset is compromised. However, if a cyber attacker can utilise the vulnerability to cause an outage of physical power system components, it is possible for a power system to be thrust into an emergency state because of the cyber-attack.

Power system cyber asset vulnerabilities are commonly understood through conventional information security principles. Successful cyber-attacks typically use some vulnerability in the communication protocol, routing or authentication processes of a cyber asset to install malware, deny legitimate services or directly intrude into an information system [39]. Successful cyber-attacks can have various consequences, but the most common are theft or corruption of information, unavailability of computing resources and physical destruction of equipment [40].

Cyber assets and applications in a power system are uniquely susceptible to various cyber-attacks. By corrupting the information utilised by a state estimator, attackers have the potential to alter the state estimation solution to reflect a fictitious power system state while evading bad data detection routines [41]. Furthermore, if an attacker can inject false data into a state estimator, they can alter the result of a security constrained optimal dispatch application, which can lead to the generator redispatching that results in uneconomic operation of a system and/or an insecure operating state [42]. Research activities have been pursued to develop more robust bad data detection algorithms that can detect a state estimator FIA [43].

Moreover, of interest is the vulnerability of Supervisory Control and Data Acquisition (SCADA) systems to cyber-attacks. While SCADA systems tend to be isolated from public communication networks, features such as remote vendor and engineering access into substation networks for legitimate purposes have created an attack surface from which a skilled attacker can obtain unauthorised entry into a SCADA system [44].

REFERENCES

1. I. Dobson, B. Carreras, V. Lynch, and D. Newman, "Complex systems analysis of series of blackouts: Cascading failure, critical points, and selforganization", *Chaos: An Interdisciplinary Journal of Nonlinear Science*, vol. 17, pp. 26–103, 2007.
2. A. Wildberger, "Complex adaptive systems: Concepts and power industry applications", *IEEE Control Systems Magazine,* vol. 17, no. 6, pp. 77–88, 2002.
3. F. Heylighen, "Five questions on complexity", Arxiv preprint online, 2007.
4. G. Chaitin, *Information, Randomness and Incompleteness: Papers on Algorithmic Information Theory*, Singapore: World Scientific Singapore, 1997.
5. S. Manson, "Simplifying complexity: A review of complexity theory", *Geoforum*, vol. 32, no. 3, pp. 405–414, 2001.
6. D.L. Harris, *M. Mitchell Waldrop: Complexity-the Emerging Science at the Edge of Order and Chaos*, New York: Simon and Schuster Inc., 1992.
7. Lewin, R., *Complejidad. El caos como generador del orden*, Barcelona: Tusquets Editores, 1995.
8. R.V. Solé and S.C. Manrubia, *Orden y caos en sistemas complejos*, Barcelona: Edicions UPC, 1996.
9. R. Solé and B.C. Goodwin, *Signs of Life: How Complexity Pervades Biology*, New York: Basic Books, 2001.
10. P. Érdi, *Complexity Explained*, Heidelberg: Springer-Verlag, 2008.
11. J. Singh, *Great Ideas in Information Theory, Language and Cybernetics*, New York: Dover Publications Inc., 1966.
12. S. Kauffman, *At Home in the Universe*, New York: Oxford University Press, 1995.
13. S. Wolfram, *A New Kind of Science*, Champaign, IL: Wolfram Media, 2002.
14. S.H. Strogatz, "Exploring complex networks", *Nature*, vol. 410, pp. 268–276, 2001.
15. V. Latora and M. Marchiori, "Efficient behaviour of small-world networks," *Physical Review Letters*, vol. 87, no. 19, p. 198701, 2001.
16. P. Erdos and A. Renyi, "On the evolution of random graphs," *Publications of the Mathematical Institute of the Hungarian Academy of Sciences*, vol. 5, pp. 17–61, 1960.
17. R. Albert, H. Jeong, and A.L. Barabasi, "Error and attack tolerance of complex networks," *Nature*, vol. 406, pp. 378–382, 2000.
18. D.J. Watts and S.H. Strogatz, "Collective dynamics of 'small-world' networks," *Nature*, vol. 393, no. 6684, pp. 440–442, 1998.
19. D.J. Watts, *Small Worlds: The Dynamics of Networks between Order and Randomness*, Princeton, NJ: Princeton University Press, pp. 11–40, 1999.
20. M. Mitchell, "Complex systems: Network thinking," *Artificial Intelligence*, vol. 170, pp. 1194–1212, 2006.

21. A.L. Barabasi and R. Albert, "Emergence of scaling in random networks," *Science*, vol. 286, no. 5439, pp. 509–512, 1999.
22. http://cs.wikipedia.org/wiki/Multiagentn%C3%AD_modelov%C3%A1n%C3 %AD.
23. L. Luo and P. Di Torino, Power system vulnerability and performance: Application from complexity scienze and complex network, Doctoral Thesis, March 2014.
24. T.A. Ernster, Power system vulnerability analysis: A centrality based approach utilizing limited information, Washington State University, Master of Science Thesis, August 2012.
25. http://web.ecs.baylor.edu/faculty/lee/ELC4340/Lecture%20note/Chapter9_GSO5.pdf.
26. H. Zhong, D. Du, C. Li, and X. Li, "A novel sparse false data injection attack method in smart grids with incomplete power network information", *Complexity*, vol. 2018, no. 2, pp. 1–16, 2018.
27. A. Dwivedi, Vulnerability analysis and fault location in power systems using complex network theory, Doctoral Thesis, RMIT University, March 2011.
28. D.J. Watts and S.H. Strogatz, "Collective dynamics of 'small-world' networks", *Nature*, vol. 393, pp. 440–442, 1998.
29. R. Albert, et al., "Error and attack tolerance of complex networks", *Nature*, vol. 406, pp. 378–382, 2000.
30. L.A.N. Amaral, et al., "Classes of small-world networks", *Proceedings of the National Academy of Sciences of the United States of America*, vol. 97, pp. 11149–11152, 2000.
31. R. Albert, et al., "Structural vulnerability of the North American power grid", *Physical Review E*, vol. 69, pp. 25103–25106, 2004.
32. M. Rosas-Casals, S. Valverde, and R. Sol_e, "Topological vulnerability of the European power grid under errors and attacks", *International Journal of Bifurcation and Chaos*, vol. 17, no. 7, pp. 2465–2475, 2007.
33. D.P. Chassin and C. Posse, "Evaluating North American electric grid reliability using the Barabasi-Albert network model", *Physica A-Statistical Mechanics and Its Applications*, vol. 355, pp. 667–677, 2005.
34. P. Crucitti, et al., "Locating critical lines in high voltage electrical power grids", *Fluctuation and Noise Letters*, vol. 5, pp. L210–L208, 2005.
35. V. Rosato, et al., "Topological properties of high-voltage electrical transmission networks", *Electric Power Systems Research*, vol. 77, pp. 99–105, 2007.
36. R. Solé, et al., "Robustness of the European power grids under intentional attack", *Physical Review E*, vol. 77, p. 26102, 2008.
37. A. Motter and Y. Lai, "Cascade-based attacks on complex networks", *Physical Review E*, vol. 66, p. 65102, 2002.
38. R. Kinney, et al., "Modeling cascading failures in the North American power grid", *European Physical Journal B*, vol. 46, pp. 101–107, 2005.
39. M. Govindarasu, A. Hann, and P. Sauer, "Cyber-physical systems security for smart grid," PSERC, Publication 12-02, February 2012.
40. C. Tranchita, N. Hadjsaid, and A. Torres, "Overview of the power systems security with regard to cyberattacks," *Fourth International Conference on Critical Infrastructures*, ISGT Latin America, Medellin, Colombia, pp. 1–8, 27 March 2009 to 30 April 2009.
41. O. Kosut, L. Jia, R.J. Thomas, and L. Tong, "On malicious data attacks on power system state estimation," *45th International Universities Power Engineering Conference*, Cornell University, USA, pp. 1–6, 31 August to 3 September 2010.
42. Y. Yuan, Z. Li, and K. Ren, "Modeling load redistribution attacks in power systems," *IEEE Transactions on Smart Grid*, vol. 2, no. 2, pp. 382–390, 2011.
43. O. Kosut, L. Jia, R.J. Thomas, and L. Tong, "Malicious data attacks on smart grid state estimation: Attack strategies and countermeasures," *First IEEE International Conference on Smart Grid Communications,* Gaithersburg, Maryland, pp. 220–225, 4–6 October 2010.

44. C. Liu, C. Ten, and M. Govindarasu, "Cybersecurity of SCADA Systems: Vulnerability assessment and mitigation," *IEEE 2009 Power Systems Conference and Exposition*, Seattle, WA, pp. 1–3, 15–18 March 2009.
45. S. Mei, X. Zhang, and M. Cao, *"Power Grid Vulnerability"*, Berlin Heidelberg: Tsinghua University Press, Beijing and Springer-Verlag, 2011.

2 Traditional Approach in Analysis of Faults in Power System

2.1 INTRODUCTION

Usually, a power system operates under balanced conditions. However, electrical fault may appear in the system leading to unbalanced system operation. Out of common electrical faults, the line fault is the most common. Much less common are the faults on cables, generators, motors and transformers. The most common faults are the single line-to-ground fault (L-G) followed by line-to-line (L-L), double line-to-ground (L-L-G) and three-phase balanced (L-L-L-G) faults.

It is important to study the system under fault condition to provide necessary system protection. In this chapter, the concept of *symmetrical components* [1] is introduced and different types of faults are analysed using symmetrical component theory so that the fault current and the voltages at different buses can be computed. Line currents have also been obtained for different fault conditions. Once the fault currents are determined at any part of the network, it becomes convenient for setting protective relays for electrical protection of power systems.

2.2 FORMATION OF [Y_{BUS}] MATRIX

Let a three-bus simple power network be considered as shown in Figure 2.1a, where Z_A and Z_B are reactances associated with EMF sources G_1 and G_2 (most commonly the alternators), and Z_C, Z_D and Z_E are reactances of the interconnecting lines between the buses 1, 2 and 3.

Let V_1, V_2 and V_3 be the respective bus voltages as shown in Figure 2.1a. Figure 2.1b is the Norton's equivalent circuit for the network shown in Figure 2.1a with EMF sources replaced by the current sources associated with corresponding admittances Y_A and Y_B. The line impedances are also replaced by corresponding admittances Y_C, Y_D and Y_E.

Application of Kirchhoff's current law (KCL) at node 1 (bus-1) yields

$$-I_1 + V_1Y_A + (V_1 - V_2)Y_D + (V_1 - V_3)Y_E = 0$$

$$\text{i.e. } I_1 = V_1Y_A + (V_1 - V_2)Y_D + (V_1 - V_3)Y_E. \tag{2.1}$$

Similarly, application of KCL at node 2 (bus-2) yields

$$I_2 = V_2Y_B + (V_2 - V_1)Y_D + (V_2 - V_3)Y_C \tag{2.2}$$

25

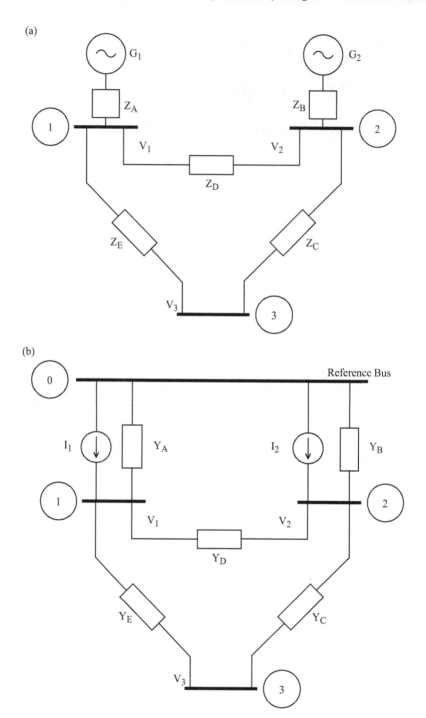

FIGURE 2.1   (a) A three-bus power network. (b) Norton equivalent circuit of the three-bus power network.

and application of KCL at node 3 (bus-3) yields,

$$0 = (V_3 - V_1)Y_E + (V_3 - V_2)Y_C \tag{2.3}$$

Rearranging Equations (2.1)–(2.3),

$$I_1 = V_1(Y_A + Y_D + Y_E) - V_2Y_D - V_3Y_E \tag{2.4}$$

$$I_2 = -V_1Y_D + V_2(Y_B + Y_C + Y_D) - V_3Y_C \tag{2.5}$$

$$0 = -V_1Y_E - V_2Y_C + V_3(Y_C + Y_E) \tag{2.6}$$

In matrix form, Equations (2.4)–(2.6) can be represented as

$$
\begin{bmatrix} I_1 \\ I_2 \\ 0 \end{bmatrix} =
\begin{bmatrix}
(Y_A + Y_D + Y_E) & -Y_D & -Y_E \\
-Y_D & (Y_B + Y_C + Y_D) & -Y_C \\
-Y_E & -Y_C & (Y_C + Y_E)
\end{bmatrix}
\begin{bmatrix} V_1 \\ V_2 \\ V_3 \end{bmatrix} \tag{2.7a}
$$

$$
\text{i.e.,} \quad
\begin{bmatrix} I_1 \\ I_2 \\ 0 \end{bmatrix} =
\begin{bmatrix}
Y_{11} & Y_{12} & Y_{13} \\
Y_{21} & Y_{22} & Y_{23} \\
Y_{31} & Y_{32} & Y_{33}
\end{bmatrix}
\begin{bmatrix} V_1 \\ V_2 \\ V_3 \end{bmatrix} \tag{2.7b}
$$

$$
\text{i.e.} \quad
\begin{bmatrix} I_1 \\ I_2 \\ 0 \end{bmatrix} = [Y_{\text{BUS}}]
\begin{bmatrix} V_1 \\ V_2 \\ V_3 \end{bmatrix}. \tag{2.7c}
$$

Here $Y_{11} = (Y_A + Y_D + Y_E)$, $Y_{12} = -Y_D$, $Y_{13} = -Y_E$, and so on.

$[Y_{\text{BUS}}]$ is called *bus admittance matrix* where Y_{11}, Y_{22} and Y_{33} are the *diagonal elements* and Y_{12}, Y_{13}, Y_{21}, Y_{23}, Y_{31} and Y_{32} are the *off-diagonal elements* of the bus admittance matrix. The diagonal elements are also termed as *self-admittances*, and each of the elements is the summation of all the admittance connected to the respective bus, while the off-diagonal elements are termed as *transfer admittances*, with each representing the admittances connected between the concerned bus and other buses.

Obviously,

$$Y_{ii} = \sum_{\substack{j=0 \\ j \neq i}}^{n} y_{ij}$$

$$\text{and,} \quad Y_{ij} = -y_{ij}$$

For the power networks, $[Y_{\text{BUS}}]$ is a square matrix of order $n \times n$, n being the number of buses. Moreover, it may be noted that $[Y_{\text{BUS}}]$ is symmetrical as $y_{ij} = y_{ji}$. Because the elements of $[Y_{BUS}]$ matrix are complex numbers in power networks, $[Y_{BUS}]$ matrix itself is complex. In addition, $Y_{ij}\,(i \neq j) = 0$ if the ith bus is not connected to bus j through a line. In real-life power system, several interconnections do not exist between a number of buses, and hence the $[Y_{BUS}]$ matrix becomes highly *sparse* (containing a number of zero elements in the matrix). This saves a lot of computer storage and memory requirements.

Example 2.1

A three-bus power network is presented in Figure 2.2. The reactances marked in the figure for the interconnecting links are in p.u. Form the $[Y_{BUS}]$ matrix.

Solution

From Figure 2.2, the line admittances can be obtained as

$$y_{12} = \frac{1}{j0.5} = -j2 \text{ p.u.} = y_{21}$$

$$y_{13} = \frac{1}{j0.8} = -j1.25 \text{ p.u.} = y_{31}$$

$$y_{23} = \frac{1}{j0.4} = -j2.5 \text{ p.u.} = y_{32}$$

$Y_{11} = y_{12} + y_{13} = -j3.25 \text{ p.u.}; \quad Y_{12} = -y_{12} = j2 \text{ p.u.} = Y_{21}; \quad Y_{13} = -y_{13} = j1.25 \text{ p.u.} = Y_{31}$

$Y_{22} = y_{21} + y_{23} = -j4.5 \text{ p.u.}; \quad Y_{23} = -y_{23} = j2.5 = Y_{32}; \quad Y_{33} = y_{31} + y_{32} = -j3.75 \text{ p.u.}$

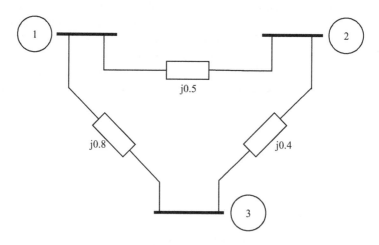

FIGURE 2.2   A three-bus power system.

$$\therefore [Y_{BUS}] = \begin{bmatrix} Y_{11} & Y_{12} & Y_{13} \\ Y_{21} & Y_{22} & Y_{23} \\ Y_{31} & Y_{32} & Y_{33} \end{bmatrix} \begin{bmatrix} -j3.25 & j2 & j1.25 \\ j & -j4.5 & j2.5 \\ j1.25 & j2.5 & -j3.75 \end{bmatrix} \text{p.u.}$$

Example 2.2

A three-bus power system is shown in Figure 2.3 indicating the p.u. line reactances of each line. Find the bus admittance matrix.

Solution

Let the ground bus be numbered as 0.

$$\therefore y_{10} = \frac{1}{Z_{10}} = \frac{1}{j0.1} = -j10 \text{ p.u.}$$

$$y_{12} = \frac{1}{Z_{12}} = \frac{1}{j0.2} = -j5 \text{ p.u.} = y_{21}$$

$$y_{23} = \frac{1}{Z_{23}} = \frac{1}{-j0.05} = j20 \text{ p.u.} = y_{32}$$

$$y_{30} = \frac{1}{Z_{30}} = \frac{1}{j0.1} = -j10 \text{ p.u.}$$

Also, $Y_{11} = y_{10} + y_{12} = -j10 - j5 = -j15$ p.u.

$$Y_{12} = -y_{12} = j5 \text{ p.u.} = Y_{21}$$

$$Y_{13} = -y_{13} = 0 = Y_{31}$$

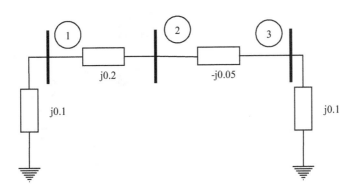

FIGURE 2.3   A three-bus power system.

$$Y_{22} = y_{20} + y_{21} + y_{23} = 0 + (-j5) + j20 = j15 \text{ p.u.}$$

$$Y_{23} = -y_{23} = -j20 \text{ p.u.} = Y_{32}$$

$$Y_{33} = y_{30} + y_{32} = -j10 + j20 = j10 \text{ p.u.}$$

$$\therefore [Y_{BUS}] = \begin{bmatrix} Y_{11} & Y_{12} & Y_{13} \\ Y_{21} & Y_{22} & Y_{23} \\ Y_{31} & Y_{32} & Y_{33} \end{bmatrix} \begin{bmatrix} -j15 & j5 & 0 \\ j5 & j15 & -j20 \\ 0 & -j20 & j10 \end{bmatrix} \text{p.u.}$$

It may be noted for this problem, in the given data, the reactance between buses 1 and 2 is ($j0.2$) p.u. while that between buses 2 and 3 is ($-j0.05$) p.u. It indicates that the reactance between buses 1 and 2 is inductive, while the reactance between buses 2 and 3 is capacitive. Moreover, there is no reactance value given between bus 2 at the reference bus. This makes y_{20}, the admittance between bus 2 and the reference bus, zero.

2.3 FORMATION OF [Y_{BUS}] WITH LINE TRANSFORMERS PRESENT

In real-life power systems, the substation is an integral part of the power network. The transmission lines are connected with transformers in the substations for ease of transmission and distribution of the electrical energy. Hence in [Y_{BUS}] formation, the transformers need to be included as an integral element with the transmission system. Similar to lines, the transformers are represented by their complex impedances along with proper representation of transformation ratio.

Figure 2.4 represents inclusion of a regulating transformer in the π model of a line connecting bus i and j. The transformer has *complex transformation ratio* $a{:}1$ ($a = |a| \angle \alpha°$). It may be noted that the π model of the line is included at the nonunity side of the transformer.

Because the transformer is assumed to have the complex off-nominal tap ratio $a{:}1$, assuming the transformer to be loss-less,

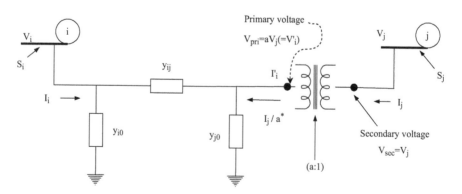

FIGURE 2.4 Representation of a regulating transformer along with the line model.

$$\frac{V_i'}{V_j} = a = |a| \angle \alpha$$

$$\text{or, } V_i'(\text{i.e., } V_{\text{pri}}) = aV_j \tag{2.8}$$

Also, when input power is equal to output power for the transformer, we have,

$$V_i' I_i^* = V_j I_j^*$$

$$\text{or, } \frac{V_i'}{V_j} = a = \frac{I_j^*}{I_i^*}$$

$\left[I_i' \text{ is the primary current of transformer while } I_j \text{ is the secondary current} \right]$

$$I_i^* = \frac{I_j}{a^*} \tag{2.9}$$

Next, we consider the current balance at two buses by the following two equations,

$$I_i = \frac{S_i^*}{V_i^*} = V_i y_{i0} + (V_i - aV_j) y_{ij} = V_i y_0 + (V_i - aV_j) y_{ij} \tag{2.10}$$

$$\text{and } \frac{I_j}{a^*} = \frac{S_i^*}{(aV_j)^*} = aV_j y_{j0} + (aV_j - V_i) y_{ij} = aV_j y_0 + (aV_j - V_i) y_{ij}$$

$$\left[\text{assuming } y_{i0} = y_{j0} = y_0 \right]$$

$$\therefore I_j = (-a^* y_{ij}) V_i + aa^* (y_0 + y_{ij}) V_j \tag{2.11}$$

Let us now rewrite Equations (2.10) and (2.11) in pair form as follows,

$$I_i = (y_0 + y_{ij}) V_i + (-ay_{ij}) V_j$$

$$I_j = (-a^* y_{ij}) V_i + aa^* (y_0 + y_{ij}) V_j$$

In matrix form, these two equations can be represented as

$$\begin{bmatrix} I_i \\ I_j \end{bmatrix} = \begin{bmatrix} y_0 + y_{ij} & -ay_{ij} \\ -ay_{ij} & aa^* (y_0 + y_{ij}) \end{bmatrix} \begin{bmatrix} V_i \\ V_j \end{bmatrix} \tag{2.12}$$

$$\text{or, } [I] = [Y][V]$$

$$\text{where, } [Y] = \begin{bmatrix} y_0 + y_{ij} & -ay_{ij} \\ -ay_{ij} & aa^*(y_0 + y_{ij}) \end{bmatrix} \tag{2.13}$$

It may be noted that if (a) is complex, then $[Y]$ is not symmetric. If (a) be a real quantity, that is, $(a) = (kV)_{base}/(kV)_{tap}$, then

$$[Y] = \begin{bmatrix} y_0 + y_{ij} & -ay_{ij} \\ -ay_{ij} & a^2(y_0 + y_{ij}) \end{bmatrix} \tag{2.14}$$

Thus, the matrix $[Y]$ becomes symmetric.

Figure 2.5 represents the inclusion of regulating transformer in the line model. Here the line model is assumed to be in the nonunity side of the complex transformation ratio (1:a).

With reference to Figure 2.5

$$\frac{V'_j}{V_i} = a = |a| \angle \alpha \tag{2.15}$$

$$\text{i.e., } V'_j = aV_i$$

$$\text{Also, } V_i I_i^* = (aV_i) I_j'^*$$

(Power is equal at the transformer input and output)

I'_j is the secondary current of transformer, while I_i is the primary current.

$$\frac{I_j'^*}{I_i^*} = \frac{1}{a}$$

$$I_i = a^* I'_j \tag{2.16}$$

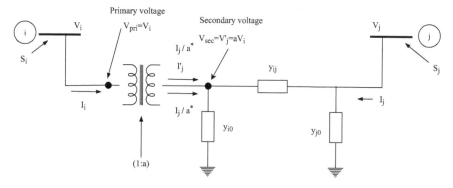

FIGURE 2.5 Inclusion of a regulating transformer with off-nominal tap ratio (1:a) in the line model.

At bus i

$$I_i = a^* I_j' = a^* \left[a V_i y_0 + \left(a V_i - V_j \right) y_{ij} \right]$$

$$\text{or, } I_i = aa^* y_0 V_i + aa^* y_{ij} V_i - a^* y_{ij} V_j$$

$$\therefore I_i = aa^* \left(y_0 + y_{ij} \right) V_i - a^* y_{ij} V_j \tag{2.17}$$

Also, at bus j

$$I_j = y_0 V_j + \left(V_j - a V_i \right) y_{ij} \quad \left[\text{assuming } y_{i0} = y_{j0} = y_0 \right]$$

$$\therefore I_j = y_0 V_j + \left(V_j - a V_i \right) y_{ij} \tag{2.18}$$

In matrix form, Equations (2.17) and (2.18) can be rearranged as

$$\begin{bmatrix} I_i \\ I_j \end{bmatrix} = \begin{bmatrix} aa^* \left(y_0 + y_{ij} \right) & -a^* y_{ij} \\ -a y_{ij} & y_0 + y_{ij} \end{bmatrix} \begin{bmatrix} V_i \\ V_j \end{bmatrix} \tag{2.19}$$

$$\therefore [Y] = \begin{bmatrix} aa^* \left(y_0 + y_{ij} \right) & -a^* y_{ij} \\ -a y_{ij} & y_0 + y_{ij} \end{bmatrix} \tag{2.20}$$

It may be noted that (a) being real,

$$[Y] = \begin{bmatrix} a^2 \left(y_0 + y_{ij} \right) & -a y_{ij} \\ -a y_{ij} & y_0 + y_{ij} \end{bmatrix} \tag{2.21}$$

the $[Y]$ matrix becomes symmetrical.

[In practical cases, the regulating transformer is designed for either *voltage magnitude* or *phase angle control*. In the former case, $\alpha = 0$ and $|a|$ can be changed in discrete steps of $\Delta|a|$. In the latter case, $|a|$ is constant and α is changed in discrete steps of $\Delta|a|$.]

In the next step $[Y_{\text{BUS}}]$, can now be modified with inclusion of the regulating transformer. Figure (2.6a) represents the π equivalent model with the regulating transformer having off-nominal tap ratio (a:1), while Figure (2.6b) represents the same with off-nominal tap ratio (1:a); the line model is always placed at the nonunity side of the transformer. It may be noted that here Y_{se}, Y_{sh1} and Y_{sh2} in respective figures are used to form Y_{12} (or Y_{21}) as well as Y_{11} and Y_{22}.

In Figure (2.6a), $Y_{se} = a y_{ij}$; $Y_{sh1} = \left(y_0 + y_{ij} \right) + \left(-a y_{ij} \right) = y_0 + \left(1 - a \right) y_{ij}$;

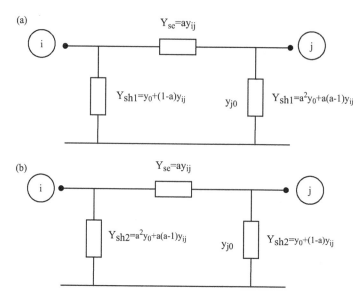

FIGURE 2.6 (a) Equivalent circuit model with transformer having off-nominal turns ratio (*a*:1). (b) Equivalent circuit model with transformer having off-nominal turns ratio (1:*a*).

$$Y_{sh2} = a^2\left(y_0 + y_{ij}\right) + \left(-ay_{ij}\right) = a^2 y_0 + a\left(a-1\right) y_{ij}$$

In Figure $(2.6b), Y_{se} = ay_{ij};\ Y_{sh1} = a^2\left(y_0 + y_{ij}\right) + \left(-ay_{ij}\right) = a^2\ y_0 + a\left(a-1\right) y_{ij}$

$$Y_{sh2} = \left(y_0 + y_{ij}\right) + \left(-ay_{ij}\right) = y_0 + \left(1-a\right) y_{ij}$$

The $[Y_{BUS}]$ matrix can be modified for inclusion of line with transformer with revised form of its self (diagonal) and transfer (off-diagonal) elements as given below:

$$Y_{ii(new)} = Y_{ii(old)} + Y_{se} + Y_{sh1}$$

$$Y_{ij(new)} = Y_{ji(new)} = Y_{ij(old)} - Y_{se} \qquad (2.22)$$

$$Y_{jj(new)} = Y_{jj(old)} + Y_{se} + Y_{sh2}$$

where $Y_{ii(old)}$, $Y_{jj(old)}$ and $Y_{ij(old)}$ are the elements of $[Y_{BUS}]$ formed without considering line transformers.

Example 2.3

A three-bus power system is shown in Figure 2.7a. Assume an ideal transformer is connected between buses 2 and 3 in series with a line reactance $j0.5$ p.u. Find $[Y_{BUS}]$. The line data for the given system is shown below (neglect shunt charging effect)

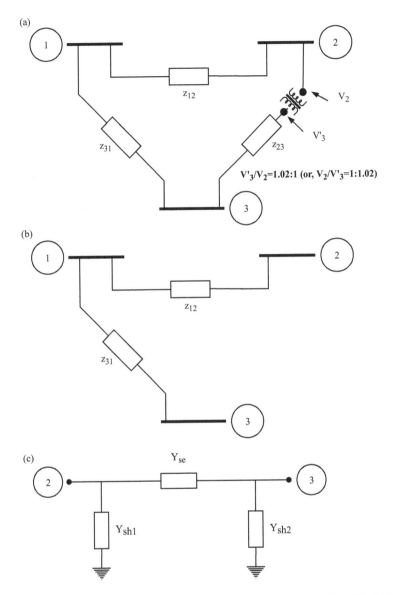

FIGURE 2.7 (a) A three-bus three-line power system with transformer in line (23). (b) System before considering the line with transformer. (c) "Pi" model of the line with transformer.

Line no.	From Bus	To Bus	R (in p.u.)	X (in p.u.)	Off-nominal Tap Ratio of Transformer
1	1	2	0.05	0.15	-
2	1	3	0.05	0.15	-
3	2	3	0	0.5	1:1.02

Solution

From the circuit of the given power system,

$$z_{12} = z_{13} = (0.05 + j0.15)\,\text{p.u.}$$

$$\therefore y_{12} = y_{13} = \frac{1}{0.05 + j0.15} = (2 - j6)\,\text{p.u.}$$

$$[Y_{\text{BUS}}] = \begin{bmatrix} (4-j12) & (-2+j6) & (-2+j6) \\ (-2+j6) & (2-j6) & (0+j0) \\ (-2+j6) & (0+j0) & (2-j6) \end{bmatrix}\text{p.u.}$$

$$\left(\because Y_{11} = y_{12} + y_{31} = (2-j6) + (2+j6) = (4-j12)\,\text{p.u.}\right)$$

With regulating transformer placed between buses 2 and 3, the $[Y_{\text{BUS}}]$ obtained needs to be modified (Figure 2.7b and c).

$$\text{Here,}\quad y_{23} = \frac{1}{j0.5} = -j2$$

$$\therefore Y_{23} = a y_{23} = 1.02 \times (-j2) = -j2.04\,\text{p.u.}$$

$$Y_{22} = a(a-1) y_{23}\,[\because y_0 = 0+j0] = 1.02 \times (1.02-1) \times (-j2) = -j0.0408\,\text{p.u.}$$

$$Y_{33} = (1-a) y_{23} = (1-1.02) \times (-j2) = j0.04\,\text{p.u.}$$

$\therefore$ For the transformer in line 2–3,

$$Y_{\text{BUS}22} = Y_{\text{BUS}22\text{old}} + (-j2.04) + (-j0.0408) = (2-j6) - j2.0808 = (2 - j8.0808)\,\text{p.u.}$$

$$Y_{\text{BUS}33} = Y_{\text{BUS}33\text{old}} + (-j2.04) + (j0.04) = (2-j6) - j2 = (-j8)\,\text{p.u.}$$

$$Y_{\text{BUS}23} = Y_{\text{BUS}32} = -(-j2.04) = j2.04\,\text{p.u.}$$

$$\therefore [Y_{\text{BUS}}]_M = \begin{bmatrix} (4-j12) & (-2+j6) & (-2+j6) \\ (-2+j6) & (2-j8.0808) & (j2.04) \\ (-2+j6) & j2.04 & (2-j8) \end{bmatrix}\text{p.u.}$$

$[Y_{\text{BUS}}]_M$ represents the modified $[Y_{\text{BUS}}]$ with the transformer included in line between buses 2 and 3.

2.4   CONCEPT OF SYMMETRICAL COMPONENT ANALYSIS

It is relatively simple to analyse a three-phase circuit in which phase voltages and currents are balanced (of equal magnitude in three phases and displaced 120° from each other), and in which the connected circuit elements are symmetrical because the treatment of a single-phase leads directly to the three-phase solution in such a case. The analysis of the circuit which is not symmetrical as a result of unbalanced load, unbalanced faults or short circuits or unbalance in supply system can effectively be conducted by "symmetrical component" analysis method first presented by C.L. Fortescue and developed by Edith Clarke.

The fundamental principle of symmetrical components as applied to the three-phase circuits is that a set of three phasors forming a three-phase unbalanced system (Figure 2.8a) can be resolved into:

a. a set of three phasors, equal in magnitude, displaced from each other by 120° in phase and having the same phase sequence as the original phasors. It forms a three-phase balanced system of *positive sequence* (Figure 2.8b).
b. a set of three phasors, equal in magnitude, displaced from each other by 120° in phase and having the phase sequence *opposite* to that of the original phasors forming a three-phase balanced system of *negative sequence* (Figure 2.8c).
c. a set of three phasors, equal in magnitude and with zero-phase displacement from each other (Figure 2.8d).

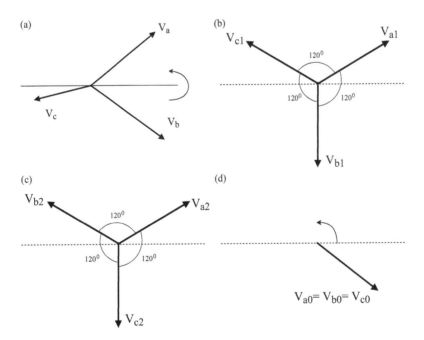

FIGURE 2.8   (a) Unbalanced phasors, (b) positive-sequence components of the phasors, (c) negative-sequence components of the phasors, and (d) zero-sequence components of the phasors.

$\therefore V_{a1}, V_{a2}, V_{a0}$ are known as the symmetrical components or, more specifically, the positive-sequence component, negative-sequence component and zero-sequence component, respectively, of V_a, and similarly V_{b1}, V_{b2}, V_{b0} for V_b and V_{c1}, V_{c2}, V_{c0} for V_c.

2.5 OPERATOR "A"

In Figure 2.9, the three phasors OP, OQ OR, are shown where each one is displaced from the other by 120°. A phasor OP′ can then be represented as OP = OP (cos θ + j sin θ) = (OP) $e^{j\theta}$, with OP being the reference phasor.

Similarly, the other two phasors can be expressed as OQ = OP $e^{j120°}$ and OR = OP $e^{j240°}$ = OP $e^{-j120°}$.

Let "a" be an operator that rotates the phasor it operates upon through 120° in a counter-clockwise direction.

Applying this definition,

$$OQ = OP \times a = a \cdot OP$$
$$OR = OP \times a \times a = a^2 \cdot OP$$

Thus, it is possible to list different values of "a" as shown below:

$$a = e^{j120°} = (\cos 120° + j\sin 120°) = (-0.5 + j0.866)$$

$$a^2 = e^{j240°} = (\cos 240° + j\sin 240°) = (-0.5 - j0.866) = e^{-j120°}$$

$$a^3 = e^{j360°} = (\cos 360° + j\sin 360°) = 1$$

$$a^4 = a^3 \times a = (-0.5 + j0.866)$$

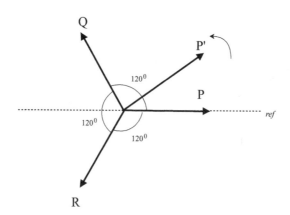

FIGURE 2.9 Rotational representation of three phasors.

$$a^5 = a^3 \times a^2 = (-0.5 - j0.866)$$

$$1 + a + a^2 = 0 = 0 + j0$$

$$a - a^2 = j1.732 = j\sqrt{3} = \sqrt{3}\angle 90°$$

$$a + a^2 = -1$$

$$1 + a = (0.5 + j0.866) = -a^2$$

$$1 - a = (1.5 - j0.866) = \sqrt{3}\angle -30°$$

2.6   SYMMETRICAL COMPONENT REPRESENTATION

Applying the concept of "a" in designating the sequence components, the positive-sequence component system (Figure 2.8b) can be expressed as

$$V_{a1} = V_{a1}; \quad V_{c1} = aV_{a1}; \quad V_{b1} = a^2 V_{a1}$$

Similarly, the symmetrical components of the negative- and zero-sequence components system can be expressed as

$$V_{a2} = V_{a2}; \quad V_{b2} = aV_{a2}; \quad V_{c2} = a^2 V_{a2}$$

$$\text{and} \quad V_{a0} = V_{a0}; \quad V_{b0} = V_{a0}; \quad V_{c0} = V_{a0}$$

Furthermore, from the concept of symmetrical component theory, the unbalanced voltages V_a, V_b and V_c can be written as phasor summation of individual symmetrical component phasors.

$$\text{i.e.} \quad V_a = V_{a1} + V_{a2} + V_{a0}$$

$$V_b = V_{b1} + V_{b2} + V_{b0} \tag{2.23}$$

$$V_c = V_{c1} + V_{c2} + V_{c0}$$

Equation (2.23) can further be represented as

$$V_a = V_{a1} + V_{a2} + V_{a0}$$

$$V_b = a^2 V_{a1} + aV_{a2} + V_{a0} \tag{2.24}$$

$$V_c = aV_{a1} + a^2 V_{a2} + V_{a0}$$

that is, in matrix form,

$$
\begin{bmatrix} V_a \\ V_b \\ V_c \end{bmatrix} = \begin{bmatrix} 1 & 1 & 1 \\ 1 & a^2 & a \\ 1 & a & a^2 \end{bmatrix} \begin{bmatrix} V_{a0} \\ V_{a1} \\ V_{a2} \end{bmatrix} = [A] \begin{bmatrix} V_{a0} \\ V_{a1} \\ V_{a2} \end{bmatrix}
\tag{2.25a}
$$

$$
\text{where } [A] = \begin{bmatrix} 1 & 1 & 1 \\ 1 & a^2 & a \\ 1 & a & a^2 \end{bmatrix}
$$

Similar to Equation (2.25a), the unbalanced phase sequence can also be represented in terms of sequence components (Equation 2.25b)

$$
\begin{bmatrix} I_a \\ I_b \\ I_c \end{bmatrix} = \begin{bmatrix} 1 & 1 & 1 \\ 1 & a^2 & a \\ 1 & a & a^2 \end{bmatrix} \begin{bmatrix} I_{a0} \\ I_{a1} \\ I_{a2} \end{bmatrix}
\tag{2.25b}
$$

Also, from Equation (2.25a), we can write

$$
\begin{bmatrix} V_{a0} \\ V_{a1} \\ V_{a2} \end{bmatrix} = [A]^{-1} \begin{bmatrix} V_a \\ V_b \\ V_c \end{bmatrix}
\tag{2.26a}
$$

$$
\text{where, } [A]^{-1} = \frac{1}{3} \begin{bmatrix} 1 & 1 & 1 \\ 1 & a & a^2 \\ 1 & a^2 & a \end{bmatrix}
$$

and from Equation (2.25b), we have

$$
\begin{bmatrix} I_{a0} \\ I_{a1} \\ I_{a2} \end{bmatrix} = [A]^{-1} \begin{bmatrix} I_a \\ I_b \\ I_c \end{bmatrix}
\tag{2.26b}
$$

2.6.1 DETERMINATION OF SEQUENCE COMPONENTS

$$
V_a + V_b + V_c = \left(1 + a^2 + a\right) V_{a1} + \left(1 + a + a^2\right) V_{a2} + 3V_{a0} = 3V_{a0} \left[\because \left(1 + a^2 + a\right) = 0 \right]
$$

$$
\therefore V_{a0} = \frac{1}{3}\left(V_a + V_b + V_c\right)
\tag{2.27}
$$

Also,

$$V_a + a^2 V_b + a V_c = V_{a1}\left(1 + a^4 + a^2\right) + V_{a2}\left(1 + a^3 + a^3\right) + V_{a0}\left(1 + a^2 + a\right) = 3V_{a2}$$

$$\therefore V_{a2} = \frac{1}{3}\left(V_a + a^2 V_b + a V_c\right) \tag{2.28}$$

and, $V_a + a V_b + a^2 V_c = V_{a1}\left(1 + a^3 + a^3\right) + V_{a2}\left(1 + a^2 + a^4\right) + V_{a0}\left(1 + a + a^2\right) = 3V_{a1}$

$$\therefore V_{a1} = \frac{1}{3}\left(V_a + a V_b + a^2 V_c\right) \tag{2.29}$$

Thus, for any unbalanced system of voltages,

$$V_{a1} = \frac{1}{3}\left(V_a + a V_b + a^2 V_c\right)$$

$$V_{a2} = \frac{1}{3}\left(V_a + a^2 V_b + a V_c\right) \tag{2.30a}$$

$$V_{a0} = \frac{1}{3}\left(V_a + V_b + V_c\right)$$

Similarly, for any unbalanced system of currents,

$$I_{a1} = \frac{1}{3}\left(I_a + a I_b + a^2 I_c\right)$$

$$I_{a2} = \frac{1}{3}\left(I_a + a^2 I_b + a I_c\right) \tag{2.30b}$$

$$I_{a0} = \frac{1}{3}\left(I_a + I_b + I_c\right)$$

From Equation (2.30a), it follows that there will be no zero-sequence component of voltage when the phasor sum $V_a + V_b + V_c = 0$. Moreover, there will be no zero-sequence component of current when the phasor sum $I_a + I_b + I_c = 0$. Because the phasor sum of the line-to-line voltages in a three-phase balanced system is always zero, there will be no zero-sequence component of the line voltages for a three-phase balanced system. Also, because the phasor sum of all the three-line currents in the three-wire three-phase system is zero, there will also be no zero-sequence component of the line current for a three-phase three-wire system.

Zero-sequence components may, however, be present in the phase voltages of a star-connected circuit or in the phase currents in a delta-connected circuit. In a star-connected circuit, the zero-sequence components of the phase voltages are all in phase. The line voltage being the phasor difference of two respective phase voltages, the zero-sequence components cancel out when line voltages are determined. In a delta-connected circuit, the zero-sequence components of the phase currents are all in phase and form a local circulating current inside the delta. Therefore, there is no zero-sequence component in the delta line currents.

2.7 CONCEPT OF SEQUENCE IMPEDANCES

The *sequence impedance* is defined as an impedance that is obtained by passing unit current of that sequence in a three-phase network or equipment and obtaining the equation for voltage drop. The sequence impedance is expressed normally in p.u. quantities. The impedance of one sequence component need not be the same to the other. Linear networks such as transmission lines normally have equal positive- and negative-sequence impedances. Transformers also have identical positive- and negative-sequence impedances, but different zero-sequence impedance. On the other hand, sequence impedances of rotating machines are the *transient* (X_d') or *sub-transient* (X_d'') reactances depending on the time of interest. Frequently, only the sequence reactance is used to calculate the fault currents and the resistance component of the impedance is neglected as its value is much less than that of the corresponding reactance. (X_d'') is normally used for calculating fault currents, while (X_d') may be used to calculate currents used to set time-delayed protection devices.

2.8 SEQUENCE COMPONENT MODELS OF PRINCIPAL POWER SYSTEM EQUIPMENT

2.8.1 GENERATOR SEQUENCE MODEL

Figure 2.10a represents the equivalent circuit of an alternator for transient conditions. Let $V_a = V_b = V_c$ and $Z_1 = Z_2 = X_d''$. The positive-sequence network contains V_a and Z_1, as shown in Figure 2.10b(i). Figure 2.10b(ii) contains only the negative sequence impedance which is mostly equal to the positive-sequence impedance.

The zero-sequence network (Figure 2.10b(iii)) includes the neutral impedance and the machine zero-sequence impedance (Z_{og}). Usually, Z_{og} varies between 0.1 and 0.7 times of (X_d''). The zero-sequence current flows in each phase of the generator and the neutral current is the sum of phase currents. Because zero-sequence current phasors have no angular offset between them (i.e. they are in phase), the neutral current is three times the current in each phase. The drop (V_{a0}) can be expressed as

$$V_{a0} = (3I_{a0})Z_n + I_{a0}Z_{og} = I_{a0}(3Z_n + Z_{og}) = I_{a0} \cdot Z_0 \qquad (2.31)$$

2.8.2 SEQUENCE NETWORK MODEL OF TRANSFORMERS

Transformers are static electrical machines and their positive- and negative-sequence impedances are the same to each other. However, the 30° phase shift in a delta star transformer needs to be considered in the sequence network models of transformers. For example, the H.V. side of the transformer is delta-connected while the L.V. side is star-connected (Figure 2.11).

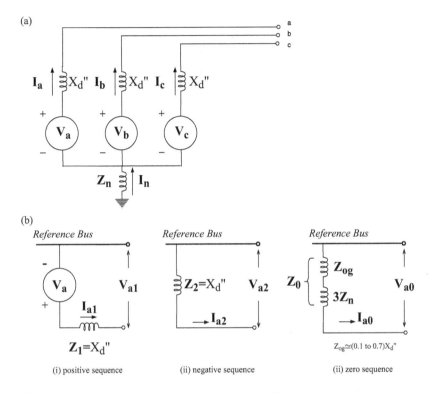

FIGURE 2.10   (a) Equivalent circuit of a generator; (b) sequence network of generator impedances: (i) positive sequence, (ii) negative sequence, (iii) zero sequence.

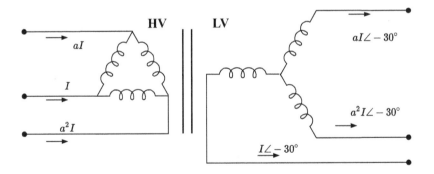

FIGURE 2.11   (−30°) Phase shift of delta star transformer.

Figure 2.11 represents the phase shift of 30° for the secondary currents. As per the ANSI standard, for positive sequence quantities for $Y-\Delta$ or $\Delta-Y$ transformers, the positive-sequence voltage to neutral on the H.V. side always leads the positive-sequence voltage to neutral on the low voltage side by 30°. Thus, when stepping up from the low-voltage side to the high-voltage side of a $\Delta-Y$ or a $Y-\Delta$ transformer, we need to advance the positive-sequence voltages (and currents) by 30° and retard the negative-sequence voltages (and currents) by 30° as the negative sequence is the reverse sequence rotation of the positive-sequence quantities.

The zero-sequence impedance of a transformer differs from the positive- and negative-sequence impedances. For an ungrounded neutral system, there is no flow of current to ground. Thus, for ungrounded star or delta-connected transformer, $Z_0 \rightarrow \infty$. Figure 2.12 represents the zero-sequence impedance of common transformer connection types.

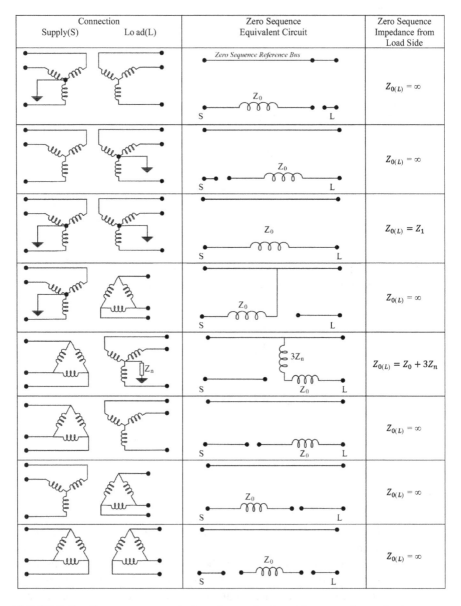

FIGURE 2.12 Zero-sequence equivalent circuits for three-phase transformer banks.

2.8.3 Sequence Impedance Circuit for Transmission Lines

The positive- and negative-sequence impedance of the transmission line (a linear network) are equal and are the given line impedances. The zero-sequence impedance is different and sometimes as a thumb rule $Z_0 = 3Z_1$.

2.8.4 Sequence Impedance Circuit for Motors

In systems where there are large motors, there is contribution of fault current by the motor. When a fault occurs, the motor connected to the system continues to rotate for a short period due to the load inertia. This rotation of the motor due to load inertia causes the motor to operate as a generator. This effect continues till the motor stops or disconnects from the system. Thus, during fault period, both the generator and the motor deliver fault current for the initial period. This effect is predominant for large motors only.

The equivalent circuit of a synchronous motor is identical to that of a synchronous generator. The positive- and negative-sequence network for the induction motor are given by the relation $(X_1) \approx (X_2) \approx (X_d) \approx 0.15 - 0.2$ p.u.

The zero-sequence impedance of the induction motor is infinite because the induction motors are not normally grounded at neutral until the motor phase voltage exceeds 4160 V. At a phase voltage beyond 4160 V, the winding may be subjected to corona effect, and hence, the motor neutral needs to be grounded. The zero-sequence impedance of a grounded neutral induction motor is not infinite, and, in such cases, the manufacturer states the value of the zero-sequence impedance.

2.9 THREE-PHASE POWER IN TERMS OF SYMMETRICAL COMPONENTS

Let S be the *complex power* flowing into a three-phase circuit through the phases r, y and b, while P and Q are the real and reactive power components of S. Let V_r, V_y and V_b be the respective phase voltages, while I_r, I_y and I_b are the corresponding three-phase currents.

$$\text{Here, } S = P + jQ = V_r I_r^* + V_y I_y^* + V_b I_b^*$$

$$\text{i.e. } S = \begin{bmatrix} V_r & V_y & V_b \end{bmatrix} \begin{bmatrix} I_r \\ I_y \\ I_b \end{bmatrix}^* = \begin{bmatrix} V_r \\ V_y \\ V_b \end{bmatrix}^T \begin{bmatrix} I_r \\ I_y \\ I_b \end{bmatrix}^* \tag{2.32}$$

However,

$$\begin{bmatrix} V_r \\ V_y \\ V_b \end{bmatrix} = [A] \begin{bmatrix} V_{r0} \\ V_{r1} \\ V_{r2} \end{bmatrix} \tag{2.33}$$

$$\text{where, } [A] = \begin{bmatrix} 1 & 1 & 1 \\ 1 & a^2 & a \\ 1 & a & a^2 \end{bmatrix}$$

$$\text{and, } \begin{vmatrix} I_r \\ I_y \\ I_b \end{vmatrix} = [A] \begin{bmatrix} I_{r0} \\ I_{r1} \\ I_{r2} \end{bmatrix} \tag{2.34}$$

Utilising Equations (2.33) and (2.34) in Equation (2.32)

$$S = [A]^T \begin{bmatrix} V_{r0} \\ V_{r1} \\ V_{r2} \end{bmatrix}^T [A]^* \begin{bmatrix} I_{r0} \\ I_{r1} \\ I_{r2} \end{bmatrix}^* \tag{2.35}$$

However, in Equation (2.35), $[A]^T [A]^* = 3u, u$ being the unity matrix.

$$[u] = \begin{bmatrix} 1 & 0 & 0 \\ 0 & 1 & 0 \\ 0 & 0 & 1 \end{bmatrix}$$

$$\therefore S = 3 \begin{bmatrix} V_{r0} & V_{r1} & V_{r2} \end{bmatrix} \begin{bmatrix} I_{r0} \\ I_{r1} \\ I_{r2} \end{bmatrix}^*$$

$$\text{or } S = 3 \left[V_{r0} I_{r0}^* + V_{r1} I_{r1}^* + V_{r2} I_{r2}^* \right] \tag{2.36}$$

that is, the sequence power is one third the power in terms of phase quantities.

Example 2.4

Figure 2.13 represents an unbalanced three-phase source having phase voltages V_a, V_b, and, V_c feeding an unbalanced load through a set of three reactances. If the phase voltages are given by $V_a = 100\angle 0°$; $V_b = 100\angle -90°$ and $V_c = 100\angle 120°$ V, find the sequence currents.

Solution

Let X_a, X_b and X_c represent the total reactances present in each phase of the given circuit.

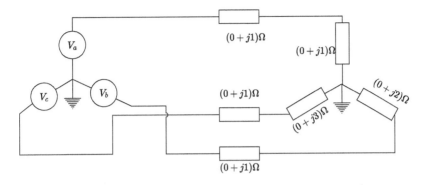

FIGURE 2.13 An unbalanced three-phase system.

$$X_a = (0+j1)+(0+j1) = j2\Omega$$

$$X_b = (0+j1)+(0+j2) = j3\Omega$$

$$X_c = (0+j1)+(0+j3) = j4\Omega$$

$$\therefore I_a = \frac{V_a}{X_a} = \frac{100\angle 0°}{j2} = -j50A = 50\angle -90°A$$

$$I_b = \frac{V_b}{X_b} = \frac{100\angle -90°}{j3} = 33.33\angle -180°A$$

$$I_c = \frac{V_c}{X_c} = \frac{100\angle 120°}{j4} = 25\angle 30°A$$

$$\therefore I_{a1} = \frac{1}{3}\left(I_a + aI_b + a^2I_c\right)$$

$$= \frac{1}{3}(50\angle -90° + 33.33\angle(-180°+120°) + 25\angle(30°+240°)$$

$$= \frac{1}{3}(50\angle -90° + 33.33\angle -60° + 25\angle 270°$$

$$= 5.55 - j34.62 = 35.06\angle -80.88°A$$

$$I_{a2} = \frac{1}{3}\left(I_a + a^2I_b + aI_c\right)$$

$$= \frac{1}{3}(50\angle -90° + 33.33\angle(-180°+240°) + 25\angle(30°+120°)$$

$$= -1.66 - j2.88 = 3.32\angle -120°A$$

$$I_{a0} = \frac{1}{3}(I_a + I_b + I_c)$$

$$= \frac{1}{3}(50\angle -90° + 33.33\angle -180° + 25\angle 30°)$$

$$= -3.89 - j12.5 = 13.09\angle -107.299° \, A$$

Example 2.5

In the circuit diagram of Figure 2.14, find the sequence currents.
    Given, $I_R = 1\angle 0°$; $I_Y = 1\angle 180°$; $I_B = 0$

Solution

$$I_{r1} = \frac{1}{3}\left(I_R + aI_Y + a^2 I_B\right)$$

$$= \frac{1}{3}\left(1\angle 0° + 1\angle(180° + 120°) + 0\right) = 0.5 - j0.2886 = 0.5773\angle -30°$$

$$I_{r2} = \frac{1}{3}\left(I_R + a^2 I_Y + aI_B\right)$$

$$= \frac{1}{3}\left(1\angle 0° + 1\angle(240° + 180°) + 0\right) = 0.5 + j0.2886 = 0.5773\angle 30°$$

$$I_{r0} = \frac{1}{3}(I_R + I_Y + I_B) = \frac{1}{3}(1\angle 0° + 1\angle 180° + 0) = 0$$

Example 2.6

In any three-phase circuit the power is denoted by the following relation

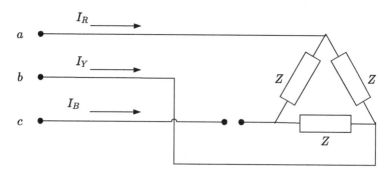

FIGURE 2.14 Circuit diagram of Example 2.5.

$$P = \begin{bmatrix} V_{AN} & V_{BN} & V_{CN} \end{bmatrix} [K] \begin{bmatrix} I_A \\ I_B \\ I_C \end{bmatrix}$$

where, V_{AN}, V_{BN} and V_{CN} are the phase voltages, and I_A, I_B and I_C are the line currents for a pure star-connected balanced load.

$$\text{If } [K] = \frac{1}{\sqrt{3}} \begin{bmatrix} 0 & 1 & -1 \\ -1 & 0 & 1 \\ 1 & -1 & 0 \end{bmatrix} \text{ and } I_A = I_B = I_C,$$

Find the magnitude of P.

Solution

$$P = \begin{bmatrix} V_{AN} & V_{BN} & V_{CN} \end{bmatrix} [K] \begin{bmatrix} I_A \\ I_B \\ I_C \end{bmatrix}$$

$$= \begin{bmatrix} V_{AN} & V_{BN} & V_{CN} \end{bmatrix} \frac{1}{\sqrt{3}} \begin{bmatrix} 0 & 1 & -1 \\ -1 & 0 & 1 \\ 1 & -1 & 0 \end{bmatrix} \begin{bmatrix} I_A \\ I_B \\ I_C \end{bmatrix}$$

$$= \left[\frac{V_{AN}}{\sqrt{3}} (I_C - I_B) + \frac{V_{BN}}{\sqrt{3}} (I_A - I_C) + \frac{V_{CN}}{\sqrt{3}} (I_B - I_A) \right]$$

$$\therefore I_A = I_B = I_C = I \text{ (say)}$$

$$P = \left[\frac{V_{AN}}{\sqrt{3}} (I - I) + \frac{V_{BN}}{\sqrt{3}} (I - I) + \frac{V_{CN}}{\sqrt{3}} (I - I) \right] = 0$$

Example 2.7

If symmetrical components of voltages and currents are given by

$$V_{a0} = 10 \angle -30° \text{ V}; \quad V_{a1} = 45 \angle 0° \text{V}; \quad V_{a2} = 25 \angle 40° \text{V}$$

and $I_{a0} = 5 \angle 190° A; \quad I_{a1} = 10 \angle 25° A; \quad I_{a2} = 1 \angle 50° A,$

Find the complex power.

Solution

$$S_{3\varnothing} = 3\left[V_{a0}I_{a0} + V_{a1}I_{a1} + V_{a2}I_{a2}\right]$$

$$= 3\left[10\angle-30°\times5\angle190° + 45\angle0°\times10\angle25° + 25\angle40°\times1\angle50°\right]$$

$$= 3\begin{bmatrix}10(\cos30° - j\sin30°)\times5(\cos190° + j\sin190°) + 45(\cos0° + j\sin0°)\\ \times10(\cos25° + j\sin25°) + 25(\cos40° + j\sin40°)\times1(\cos50° + j\sin50°)\end{bmatrix}$$

$$= 3\left[360.85 + j232.28\right]$$

$$= 1082.56 + j696.84 = 1287.448\angle32.77°$$

2.10   [Z_{BUS}] BUILDING (STEP-BY-STEP METHOD)

During the formulation of [Y_{BUS}], [Z_{BUS}] can be obtained by inversion of [Y_{BUS}]. However, this technique may consume more computer time and memory once the size of the system is large and the [Y_{BUS}] matrix is extremely sparse. Moreover, any modification in the network needs reformation of [Y_{BUS}], and for large systems this is clearly not convenient.

On the other hand, the process of [Z_{BUS}] *building* with step-by-step method is more convenient and the technique starts working with branch impedance values. The algorithm starts from scratch and any modification in the network does not require a complete rebuilding of [Z_{BUS}].

Once the reference node is defined and the network bus voltage is specified with respect to this node, a branch (Z_b say) can be easily added to the existing impedance matrix [Z_{BUS}]$_{old}$ to produce the new impedance matrix. There are five cases where the addition of the branch can influence the [Z_{BUS}], depending on the type of Z_b with system buses (i.e. whether with reference bus, new bus or old buses).

2.10.1   ADDING A BRANCH (OR LINK) Z_B FROM A NEW BUS TO THE REFERENCE BUS (TYPE 1 MODIFICATION)

The addition of new bus (k) to the reference bus (or node) through a link/branch of impedance Z_b (Figure 2.15) without any connection to any of the other buses of the original network will not alter the original bus voltages, even if a new current I_k is injected in the new bus (k). The voltage V_k at the new bus is given by

$$V_k = Z_b I_k$$

Obviously, $Z_{ik} = Z_{ki} = 0;\quad i = 1,2,3,\ldots,n$ $\qquad(2.37)$

$$Z_{kk} = Z_b$$

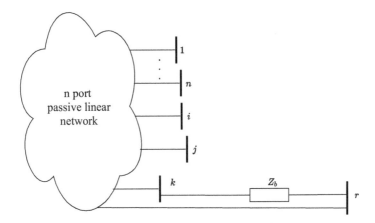

FIGURE 2.15   Addition of link Z_b from new bus (k) to an old bus (r).

Then, the nodal equation becomes

$$
\begin{bmatrix} V_1 \\ V_2 \\ \vdots \\ V_n \\ \cdots \\ V_k \end{bmatrix}
=
\left[
\begin{array}{ccccc|c}
 & & & & & 0 \\
 & & [Z_{\text{BUS}}]_{\text{old}} & & & \vdots \\
 & & & & & \vdots \\
 & & & & & 0 \\
\hline
0 & \cdots & \cdots & 0 & | & Z_b
\end{array}
\right]
\begin{bmatrix} I_1 \\ I_2 \\ \vdots \\ I_n \\ \cdots \\ I_k \end{bmatrix}
\qquad (2.38a)
$$

$$
\text{or } [V] = [Z_{\text{BUS}}]_{\text{new}} [I]
$$

$$
\text{Here, } [Z_{\text{BUS}}]_{\text{new}} =
\left[
\begin{array}{ccccc|c}
 & & & & & 0 \\
 & & [Z_{\text{BUS}}]_{\text{old}} & & & \vdots \\
 & & & & & \vdots \\
 & & & & & 0 \\
\hline
0 & \cdots & \cdots & 0 & | & Z_b
\end{array}
\right]
\qquad (2.38b)
$$

It may be noted here that the column vector of the current multiplied by $[Z_{\text{BUS}}]_{\text{new}}$ does not alter the voltage of the original network.

2.10.2 ADDITION OF A BRANCH (OR LINK) Z_B FROM A NEW BUS TO AN OLD BUS (TYPE 2 MODIFICATION)

Let a branch of impedance Z_b be added from a new bus (k) to an old bus (j). With reference to Figure 2.16 the current I_k entering into bus j would increase the original voltage (V_j) of the bus j by a voltage $I_k Z_b$, that is,

$$V_k = V_j + Z_b I_k$$

$$= Z_{j1} I_1 + Z_{j2} I_2 + \cdots + Z_{jj}\left(I_j + I_k\right) + \cdots + Z_{jn} I_n + Z_b I_k \qquad (2.39)$$

$$\text{or } V_k = Z_{j1} I_1 + Z_{j2} I_2 + \cdots + Z_{jj} I_j + \cdots + Z_{jn} I_n + \left(Z_{jj} + Z_b\right) I_k$$

Thus, the new row must be combined with $[Z_{\text{BUS}}]_{\text{old}}$ to find V_k; this is $\{Z_{j1}\ Z_{j2}\ ...Z_{jn}\ (Z_{jj}+Z_b)\}$. Because $[Z_{\text{BUS}}]$ is to be a square matrix around the principal diagonal, a new column is to be combined that is a transpose of the new row. The new column accounts for the increase of all bus voltages due to I_k. The generalised matrix equation then becomes

$$
\begin{bmatrix}
V_1 \\
V_2 \\
\vdots \\
V_n \\
\cdots \\
V_k
\end{bmatrix}
=
\begin{bmatrix}
 & & & & | & Z_{1j} \\
 & [Z_{\text{BUS}}]_{\text{old}} & & & | & Z_{2j} \\
 & & & & | & \vdots \\
 & & & & | & Z_{nj} \\
\hline
Z_{j1} & Z_{j2} & \cdots & Z_{jn} & | & \left(Z_{jj} + Z_b\right)
\end{bmatrix}
\begin{bmatrix}
I_1 \\
I_2 \\
\vdots \\
I_n \\
\cdots \\
I_k
\end{bmatrix}
\qquad (2.40a)
$$

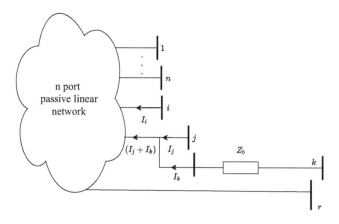

FIGURE 2.16　Addition of a link Z_b from new bus (k) to an old bus (j).

$$\text{or, } [V] = [Z_{BUS}]_{new} [I]$$

$$\text{Here, } [Z_{BUS}]_{new} = \begin{bmatrix} & & & & | & Z_{1j} \\ & [Z_{BUS}]_{old} & & & | & Z_{2j} \\ & & & & | & \vdots \\ & & & & | & Z_{nj} \\ - & - & - & - & | & - \\ Z_{j1} & Z_{j2} & \cdots & Z_{jn} & | & (Z_{jj}+Z_b) \end{bmatrix} \quad (2.40b)$$

[It may be observed here that the first elements of the new row of $[Z_{BUS}]_{new}$ are the elements of row j of $[Z_{BUS}]_{old}$, and the first n elements of the new column are the elements of column j of $[Z_{BUS}]_{old}$]

2.10.3 ADDITION OF A BRANCH (OR LINK) Z_B FROM AN OLD BUS TO THE REFERENCE BUS (TYPE 3 MODIFICATION)

Let there be an addition of a branch (link) Z_b from an old bus j to the reference bus (Figure 2.17). This is done by short circuiting bus (k) to reference node (r) by making $V_k = 0$. This will yield the same matrix as shown in Equation (2.40a) except for $V_k = 0$.

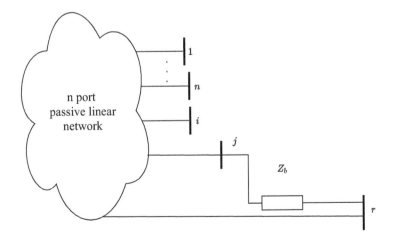

FIGURE 2.17 Addition of a link (Z_b) from an old bus (j) to the reference bus (r).

This yields

$$
\begin{bmatrix} V_1 \\ V_2 \\ \vdots \\ V_n \\ \cdots \\ 0 \end{bmatrix} = \left[\begin{array}{cccc|c} & & & & Z_{1j} \\ & [Z_{\text{BUS}}]_{\text{old}} & & & Z_{2j} \\ & & & & \vdots \\ & & & & Z_{nj} \\ \hline Z_{j1} & Z_{j2} & \cdots & Z_{jn} & (Z_{jj}+Z_b) \end{array} \right] \begin{bmatrix} I_1 \\ I_2 \\ \vdots \\ I_n \\ \cdots \\ I_k \end{bmatrix} \tag{2.41}
$$

A new $(n+1)$th row and a new $(n+1)$th column is obtained. It is now possible to eliminate $(n+1)$th row and $(n+1)$th column by Kron reduction. Each element in the new matrix will then become

$$
Z_{m(\text{new})} = Z_{mi} - \frac{Z_{m(n+1)}Z_{(n+1)i}}{Z_{jj}+Z_b} \tag{2.42}
$$

or, $[Z_{\text{BUS}}]_{\text{new}} = [Z_{\text{BUS}}]_{\text{old}}$

$$
- \frac{1}{Z_{jj}+Z_b} \begin{bmatrix} Z_{1j} \\ \vdots \\ Z_{nj} \end{bmatrix} \begin{bmatrix} Z_{j1} & \cdots & Z_{jn} \end{bmatrix} \tag{2.43}
$$

2.10.4 ADDITION OF A BRANCH (OR LINK) Z_B BETWEEN TWO OLD BUSES (TYPE 4 MODIFICATION)

Let a branch (link) Z_b be added from an old bus i to another old bus j (Figure 2.18) For bus 1,

$$
V_1 = Z_{11}I_1 + Z_{12}I_2 + \cdots + Z_{1i}(I_i+I_k) + Z_{1j}(I_j-I_k) + \cdots + Z_{1n}I_n \tag{2.44}
$$

Rearrangement of Equation (2.44) yields

$$
V_1 = Z_{11}I_1 + Z_{12}I_2 + \cdots + (Z_{1i}-Z_{1j})I_k + \cdots + Z_{1n}I_n \tag{2.45}
$$

Similar equation can be written for all other buses.

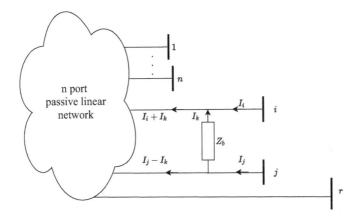

FIGURE 2.18 Addition of a link (Z_b) between two old buses (i and j).

Finally,

$$V_j = V_i + Z_b I_k$$

$$= Z_{j1}I_1 + Z_{j2}I_2 + \cdots + Z_{ji}\left(I_i + I_k\right) + Z_{jj}\left(I_j - I_k\right) + \cdots + Z_{jn}I_n \qquad (2.46)$$

$$= Z_{i1}I_1 + Z_{i2}I_2 + \cdots + Z_{ii}\left(I_i + I_k\right) + Z_{ij}\left(I_j - I_k\right) + \cdots + Z_{in}I_n + Z_b I_k$$

Rearranging Equation (2.45),

$$0 = \left(Z_{i1} - Z_{j1}\right)I_1 + \cdots + \left(Z_{ii} - Z_{ji}\right)I_i + \left(Z_{ij} - Z_{jj}\right)I_j + \cdots$$

$$+ \left(Z_{in} - Z_{jn}\right)I_n + \left(Z_b + Z_{ii} + Z_{jj} - Z_{ji} - Z_{ij}\right)I_k$$

In a matrix form,

$$
\begin{bmatrix} V_1 \\ \vdots \\ V_n \\ \cdots \\ 0 \end{bmatrix}
=
\left[
\begin{array}{cccc}
 & & | & \left(Z_{1i} - Z_{1j}\right) \\
 & \left[Z_{\text{BUS}}\right]_{\text{old}} & | & \vdots \\
 & & | & \left(Z_{ni} - Z_{nj}\right) \\
\cline{1-4}
\left(Z_{i1} - Z_{j1}\right) & \cdots \quad \cdots & | & \left(Z_b + Z_{ii} + Z_{jj} - 2Z_{ij}\right)
\end{array}
\right]
\begin{bmatrix} I_1 \\ \vdots \\ I_n \\ \cdots \\ I_k \end{bmatrix}
$$

$$(2.47)$$

Elimination of I_k yields Equation (2.48)

$$[Z_{BUS}]_{new} = [Z_{BUS}]_{old} - \frac{1}{(Z_b + Z_{ii} + Z_{jj} - 2Z_{ij})} \begin{bmatrix} Z_{1i} - Z_{1j} \\ \vdots \\ Z_{ni} - Z_{nj} \end{bmatrix}$$

$$\times \begin{bmatrix} (Z_{i1} - Z_{j1}) & \cdots & (Z_{in} - Z_{jn}) \end{bmatrix}$$

(2.48)

2.10.5 ADDITION OF TWO BRANCHES Z_A AND Z_B WITH MUTUAL IMPEDANCE (Z_M) BETWEEN FOUR BUSES (TYPE 5 MODIFICATION)

Let there be two branches Z_a and Z_b added with mutual impedance Z_m between four buses j, k, l, and m, as shown in Figure 2.19. Here,

$$V_i = Z_{i1}I_1 + Z_{i2}I_2 + \cdots + Z_{ij}(I_j - I_a) + Z_{ik}(I_k + I_a)$$

$$+ Z_{il}(I_l - I_b) + Z_{im}(I_m + I_b) + \cdots + Z_{in}I_n$$

(2.49)

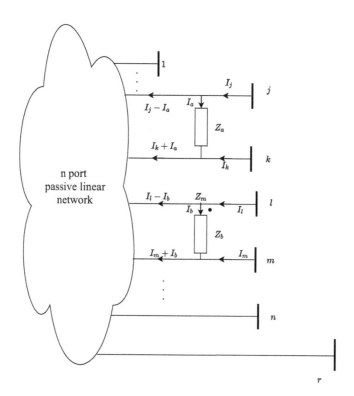

FIGURE 2.19 Addition of two branches (Z_a and Z_b) with mutual impedance (Z_m) between four buses.

On rearrangement, Equation (2.49) becomes

$$V_i = Z_{i1}I_1 + \cdots + Z_{ij}I_j + Z_{ik}I_k + Z_{il}I_l + Z_{im}I_m + \cdots$$
$$+ Z_{in}I_n + \left(Z_{ik} - Z_{ij}\right)I_a + \left(Z_{im} - Z_{il}\right)I_b \qquad (2.50)$$

Again, at the j-th and k-th buses, with mutual impedance Z_m between them V_j is given by

$$V_j = Z_a I_a + Z_m I_b + V_k \qquad (2.51)$$

Equation (2.50) can be used in Equation (2.51) with proper notations (for j-th and k-th buses) to yield

$$Z_{j1}I_1 + Z_{j2}I_2 + \cdots + Z_{jn}I_n + \left(Z_{jk} - Z_{jj}\right)I_a + \left(Z_{jm} - Z_{jl}\right)I_b$$
$$= Z_a I_a + Z_m I_b + Z_{k1}I_1 + \cdots + Z_{kn}I_n + \left(Z_{kk} - Z_{kj}\right)I_a + \left(Z_{km} - Z_{kl}\right)I_b \qquad (2.52)$$

On rearrangement, Equation (2.52) becomes

$$0 = \left(Z_{k1} - Z_{j1}\right)I_1 + \cdots + \left(Z_{kn} - Z_{jn}\right)I_n + \left(Z_a + Z_{jj} + Z_{kk} - 2Z_{jk}\right)I_a$$
$$+ \left(Z_m + Z_{jl} + Z_{km} - Z_{jm} - Z_{kl}\right)I_b \qquad (2.53)$$

Again, for buses l and m,

$$V_l = Z_b I_b + Z_m I_a + V_m$$

or, $$Z_{l1}I_1 + \cdots + Z_{ln}I_n + \left(Z_{lk} - Z_{lj}\right)I_a + \left(Z_{lm} - Z_{ll}\right)I_b$$
$$= Z_b I_b + Z_m I_a + Z_{m1}I_1 + \cdots + Z_{mn}I_n + \left(Z_{mk} - Z_{mj}\right)I_a + \left(Z_{mm} - Z_{ml}\right)I_b \qquad (2.54)$$

On rearrangement, Equation (2.54) becomes

$$0 = \left(Z_{m1} - Z_{l1}\right)I_1 + \cdots + \left(Z_{mn} - Z_{ln}\right)I_n + \left(Z_m + Z_{lj} + Z_{mk} - Z_{lk} - Z_{mj}\right)I_a$$
$$+ \left(Z_b + Z_{ll} + Z_{mn} - 2Z_{lm}\right)I_b \qquad (2.55)$$

Thus, in matrix representation, Equation (2.55) can be represented as

$$
\begin{bmatrix} V_1 \\ \vdots \\ V_n \\ -- \\ 0 \\ \vdots \\ 0 \end{bmatrix} = \begin{bmatrix} [Z_{BUS}]_{old} & \vline & (Z_{lk} - Z_{ij}) & \vline & (Z_{im} - Z_{il}) \\ - & \vline & - & - & - \\ \cdots(Z_{ki} - Z_{ji})\cdots & \vline & (Z_a + Z_{ij} + Z_{kk} - 2Z_{jk}) & \vline & (Z_m + Z_{jl} + Z_{km} - Z_{kl} - Z_{jm}) \\ \cdots(Z_{mi} - Z_{li})\cdots & \vline & (Z_m + Z_{lj} + Z_{mk} - Z_{lk} - Z_{mj}) & \vline & (Z_b + Z_{ll} + Z_{mm} - 2Z_{lm}) \end{bmatrix}
$$

$$
\times \begin{bmatrix} I_1 \\ \vdots \\ I_n \\ -- \\ I_a \\ -- \\ I_b \end{bmatrix}
$$

$$
\text{i.e., } [Z_{BUS}]_{new} = \begin{bmatrix} [Z_{BUS}]_{old} & [Z_{ab}] \\ [Z_{ba}] & [Z_{bb}] \end{bmatrix} \tag{2.56}
$$

$$
= [Z_{BUS}]_{old} - [Z_{ab}][Z_{bb}]^{-1}[Z_{ba}]
$$

The algorithm to obtain bus impedance matrix $[Z_{BUS}]$ is given in the next article.

2.11   ALGORITHM FOR FORMATION OF BUS IMPEDANCE MATRIX [Z_{BUS}] USING STEP-BY-STEP METHOD

1. Read existing $[Z_{BUS}]$ matrix, that is, $[Z_{BUS}]_{old}$.
2. Read modification type i.
3. Read the added impedance value Z_b.
4. If $i = 1$ then obtain $[Z_{BUS}]_{new}$ using Equation (2.38b) and go to step 9, else go to the next step.
5. If $i = 2$ then obtain $[Z_{BUS}]_{new}$ using Equation (2.40b) and go to step 9, else go to the next step.
6. If $i = 3$ obtain $[Z_{BUS}]_{new}$ using Equation (2.40a) and go to step 9, else go to the next step.
7. If $i = 4$ then obtain $[Z_{BUS}]_{new}$ using Equation (2.48) and go to step 9, else go to the next step.
8. If $i = 5$ obtain $[Z_{BUS}]_{new}$ using Equation (2.56a), else go to the next step.
9. Go to step 2 for further modification, else go to the next step.
10. Display and/or store final $[Z_{BUS}]$.
11. Stop.

Example 2.8

Figure 2.20 shows a four-bus power network. Assuming bus 1 to be the reference bus, find $[Z_{BUS}]$.

Solution

Bus 1 is the reference bus and in the algorithm of $[Z_{BUS}]$ building, the bus numbering is done excluding the reference bus. Hence, new bus numbering is required as given below.

Old Bus no.	New Bus no.
1	Reference (r)
2	1
3	2
4	3

Figure 2.21 represents the redesignated four-bus network.

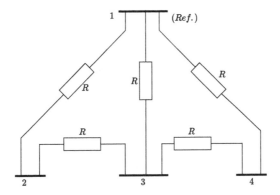

FIGURE 2.20 A four-bus network ($R = 5\ \Omega$).

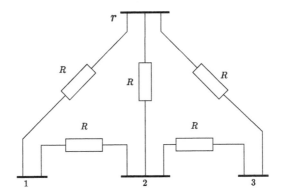

FIGURE 2.21 Redesignated four-bus network.

Initially, order of old $[Z_{BUS}]$ is assumed to be zero.

Step 1: Consider the element between bus 1 and reference $[Z_b = 5\Omega]$ (type 1 modification) where added impedance is 5 Ω.

∵ Added impedance value = 5Ω,

$$[Z_{Bus}] = \begin{matrix} & 1 \\ 1 & [5] \end{matrix} \ \Omega$$

Step 2: In the next modification consider the element between new bus 2 and reference bus $[Z_b = 5\Omega]$. This is also type 1 modification.

$$[Z_{Bus}] = \begin{matrix} & & 1 & 2 \\ 1 & \begin{bmatrix} 5 & 0 \\ 0 & 5 \end{bmatrix} \end{matrix} \ \Omega$$

Step 3: Consider the element $[Z_b = 5\Omega]$ between bus 3 and reference bus. This is again a type 1 modification between the new bus and reference bus.

$$\therefore [Z_{Bus}] = \begin{matrix} & & 1 & 2 & 3 \\ 1 & \begin{bmatrix} 5 & 0 & 0 \\ 0 & 5 & 0 \\ 0 & 0 & 5 \end{bmatrix} \end{matrix} \ \Omega$$

Step 4: Consider the element $[Z_b = 5\Omega]$ between buses 1 and 2 in the next modification. This is type 4 modification between the two old buses 1 and 2.

$$[Z_{Bus}] = [Z_{BUS}]_{old} - \frac{1}{5 + Z_{11} + Z_{22} - 2Z_{12}} \begin{bmatrix} (Z_{11} - Z_{12}) \\ (Z_{21} - Z_{22}) \\ (Z_{31} - Z_{32}) \end{bmatrix} \times (Z_{11} - Z_{21})(Z_{12} - Z_{22})(Z_{13} - Z_{23})$$

$$\text{or, } [Z_{Bus}] = [Z_{BUS}]_{old} - \frac{1}{5 + 5 + 5} \begin{bmatrix} 5 \\ -5 \\ 0 \end{bmatrix} \begin{bmatrix} 5 & -5 & 0 \end{bmatrix}$$

$$\text{or, } [Z_{Bus}] = \begin{bmatrix} 5 & 0 & 0 \\ 0 & 5 & 0 \\ 0 & 0 & 5 \end{bmatrix} - \frac{1}{15} \begin{bmatrix} 5 \\ -5 \\ 0 \end{bmatrix} \begin{bmatrix} 25 & -25 & 0 \\ -25 & 25 & 0 \\ 0 & 0 & 0 \end{bmatrix}$$

Simplification yields

$$\text{or, } [Z_{Bus}] = \begin{matrix} & & 1 & 2 & 3 \\ 1 & \begin{bmatrix} 3.33 & 1.67 & 0 \\ 1.67 & 3.33 & 0 \\ 0 & 0 & 5 \end{bmatrix} \end{matrix} \ \Omega$$

Step 5: In the next modification, consider the element $[Z_b = 5\Omega]$ between buses 2 and 3. This is also type 4 modification.

Added impedance value is 5 Ω between two old buses 2 and 3.

$$[Z_{Bus}] = [Z_{BUS}]_{old} - \frac{1}{5 + Z_{22} + Z_{33} - 2Z_{23}} \times \begin{bmatrix} (Z_{12} - Z_{13}) \\ (Z_{22} - Z_{23}) \\ (Z_{32} - Z_{33}) \end{bmatrix} (Z_{21} - Z_{31})(Z_{22} - Z_{32})(Z_{23} - Z_{33})$$

$$= [Z_{BUS}]_{old} - \frac{1}{5 + 3.33 + 5} \begin{bmatrix} 1.67 \\ 3.33 \\ -5 \end{bmatrix} \begin{bmatrix} 1.67 & 3.33 & -5 \end{bmatrix}$$

$$= [Z_{BUS}]_{old} - \frac{1}{13.33} \begin{bmatrix} 1.67 \\ 3.33 \\ -5 \end{bmatrix} \begin{bmatrix} 1.67 & 3.33 & -5 \end{bmatrix}$$

$$= \begin{bmatrix} 3.33 & 1.67 & 0 \\ 1.67 & 3.33 & 0 \\ 0 & 0 & 5 \end{bmatrix} - \begin{bmatrix} 0.21 & 0.42 & -0.62 \\ 0.42 & 0.83 & -1.25 \\ -0.62 & -1.25 & 1.87 \end{bmatrix}$$

$$= \begin{matrix} & 1 & 2 & 3 \\ 1 \\ 2 \\ 3 \end{matrix} \begin{bmatrix} 3.12 & 1.25 & 0.62 \\ 1.25 & 2.5 & 1.25 \\ 0.62 & 1.25 & 3.12 \end{bmatrix} \Omega$$

Because there is no further modification and no bus interchange, the above matrix is the final $[Z_{Bus}]$ matrix.

Therefore,

$$[Z_{Bus}] = \begin{matrix} & 1 & 2 & 3 \\ 1 \\ 2 \\ 3 \end{matrix} \begin{bmatrix} 3.12 & 1.25 & 0.62 \\ 1.25 & 2.5 & 1.25 \\ 0.62 & 1.25 & 3.12 \end{bmatrix} \Omega = [Z_{Bus}]_{new}$$

2.12 DETERMINATION OF SYMMETRICAL FAULT CURRENT USING $[Z_{BUS}]$ INVERSION

A simple way to find symmetrical fault current at any bus of multi-bus power system is to use $[Z_{Bus}]$ obtained directly by inverting $[Y_{Bus}]$ as

$$[Z_{Bus}] = [Y_{Bus}]^{-1} = \begin{bmatrix} Y_{11} & Y_{12} & \cdots & Y_{1i} & \cdots & Y_{1n} \\ Y_{21} & Y_{22} & \cdots & Y_{2i} & \cdots & Y_{2n} \\ \vdots & \vdots & & \vdots & & \vdots \\ Y_{i1} & Y_{i2} & \cdots & Y_{ii} & \cdots & Y_{in} \\ \vdots & \vdots & & \vdots & & \vdots \\ Y_{n1} & Y_{n2} & \cdots & Y_{ni} & \cdots & Y_{nn} \end{bmatrix}^{-1}$$

$$= \begin{bmatrix} Z_{11} & Z_{12} & \cdots & Z_{1i} & \cdots & Z_{1n} \\ Z_{21} & Z_{22} & \cdots & Z_{2i} & \cdots & Z_{2n} \\ \vdots & \vdots & & \vdots & & \vdots \\ Z_{i1} & Z_{i2} & \cdots & Z_{ii} & \cdots & Z_{in} \\ \vdots & \vdots & & \vdots & & \vdots \\ Z_{n1} & Z_{n2} & \cdots & Z_{ni} & \cdots & Z_{nn} \end{bmatrix}$$

(2.57)

This symmetrical fault current at bus 1 (say) is given by

$I_{f1} = \frac{1}{Z_{11}}$, assuming the pre-fault bus voltage of bus 1 as 1.00 p.u. Similarly, the symmetrical fault current of any bus i is given by $I_{fi} = \frac{1}{Z_{ii}}$, assuming the pre-fault i th bus voltage to be 1.00. However, for large multi-bus power network, this approach of finding the symmetrical bus fault current is unacceptable because of the need of inverting a very large and sparse $[Y_{Bus}]$ matrix. Furthermore, for any change in the network, the full $[Y_{Bus}]$ is to be rebuilt, and again it needs to be inverted using efficient $[Y_{Bus}]$ inversion technique. On the other hand, finding fault current from $[Z_{Bus}]$ where the direct building of $[Z_{Bus}]$ algorithm is used appears to be more reasonable and saves computer time and memory.

2.13   PHASE-SEQUENCE COMPONENT NETWORK

It has already been established that using the symmetrical component theory, it is possible to replace a three-phase set of unbalanced voltages (or currents) defined as *positive-phase sequence (PPS)*, *negative-phase sequence (NPS)* and *zero-phase sequence (ZPS)* sets. Because of an external unbalanced condition, say a single-phase short circuit, PPS, NPS and ZPS voltages and currents appear on the network at a fault point. A phase-sequence component is one that carries current and voltages of one particular phase sequence such as PPS, NPS or ZPS sequence. It may be noted that because the actual three-phase network is assumed to be balanced, PPS, NPS and ZPS networks are separate and there is no intersequence mutual coupling between them. These sequence networks are only connected at the point of unbalance in the system. Moreover, the assumption of a perfectly balanced three-phase network means that only positive sequence voltage and currents exist in the PPS network. Although negative and ZPS network can still be artificially constructed, they are totally redundant because they carry no negative or ZPS voltages or currents. The negative- and zero-sequence voltages in the corresponding negative and ZPS networks appear as a result of the unbalanced condition imposed on the actual three-phase network; they should not be confused with the voltage sources that exist in the PPS network.

Let the unbalanced condition be considered for a three-phase network at point F, as shown in Figure 2.22a. The three-phase networks can be then constructed from the actual three-phase network components and network theory and are shown in Figures 2.22b and c. The entire PPS, NPS and ZPS networks are derived using Thevenin's theorem. This reduction results in a single equivalent voltage source at point F and a single equivalent impedance seen looking back into the relevant network from point F, as illustrated in Figure 2.22c.

From Figure 2.22c, sequence voltage and current relations at the point of fault (F) can be written for the networks. For the active positive-sequence network,

$$V_n = V_F - Z_1 I_n \tag{2.58}$$

where V_F is the positive-sequence phase voltage at the point of fault immediately before the event of unbalance appearing at F. V_n is the resultant positive sequence voltage at the point of fault, and I_n is the flow of positive sequence current in the positive sequence network.

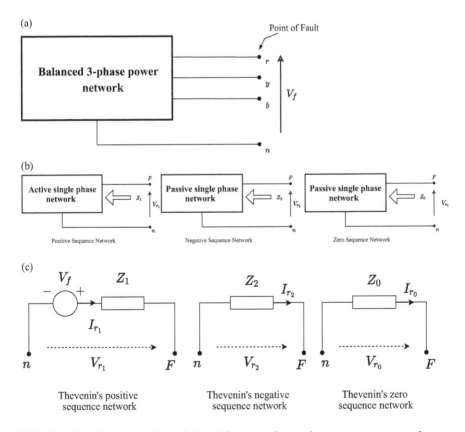

FIGURE 2.22 Equivalent Thevenin's positive-, negative- and zero-sequence networks.

For the passive negative and zero-sequence networks,

$$V_{r2} = -Z_2 I_{r2} \tag{2.59}$$

$$V_{r0} = -Z_0 I_{r0} \tag{2.60}$$

where V_{r2} and V_{r0} are the negative-sequence and zero-sequence voltages at F for the negative- and zero-sequence networks, respectively. I_{r2} and I_{r0} are the corresponding negative- and zero-sequence currents flowing out of the NPS and zero-sequence networks into the fault, respectively. Z_1, Z_2 and Z_0 represent the positive-, negative- and zero-sequence impedances of the sequence networks.

In the multi-bus power system, the governing sequence network equation can be represented, generalising Equations (2.58)–(2.60) as follows:

For the positive-sequence generalised network,

$$V_{i_1} = E - \sum_{k=1}^{n} Z_{ik_1} I_{k_1} \tag{2.61a}$$

where V_{i_1} is the generalised-positive sequence voltage at ith bus, the bus subjected to be under fault condition, and E is the pre-fault bus phase voltage. I_{k_1} is the positive-sequence current at any bus k, while Z_{ik_1} is the equivalent positive-sequence impedance looking into the positive-sequence network of bus k. For the convenience of the modelling, it is the logical assumption that all the currents at bus $k \neq i$ are zero, while I_f is the fault current at bus i at $k = i$. Hence, I_{k_1}, I_{k_2} and I_{k_0} are all zero quantities for $k \neq i$. At bus, the positive-, negative- and zero-sequence currents are represented as I_{i_1}, I_{i_2} and I_{i_0} $(k = i)$.

Then, for the faulted power network, the governing equation of the sequence networks are given by

$$V_{i_1} = E - [Z_1] I_{i_1} \tag{2.61b}$$

$$V_{i_2} = -\sum_{k=1}^{n} Z_{ik_2} I_{k_2} \tag{2.62}$$

$$V_{i_0} = -\sum_{k=1}^{n} Z_{ik_0} I_{k_0} \tag{2.63}$$

$$= -[Z_0] I_{i_0}$$

The following common types of electrical faults are analysed in the following articles:

a. Three-phase balanced fault
b. Single line-to-ground fault
c. Line-to-line fault
d. Line-to-line-to-ground fault

For any of the above type of fault at the i-th bus, $I_k = 0$ $(k = 1, 2, 3, \ldots, n; k \neq i)$.

2.13.1 THREE-PHASE BALANCED FAULT

A balanced three-phase system remains symmetrical after the occurrence of a three-phase fault having the same impedance between each line and a common point. Only positive-sequence current flows through such a network. The positive-sequence voltages at the i-th bus, where the three-phase balanced (symmetrical) fault appears is rewritten from Equation (2.61).

$$V_{i(1)} = E - \sum_{k=1}^{n} Z_{ik(1)} I_{k(1)}$$

$$= E - \left[Z_{i1(1)} I_{1(1)} + \cdots + Z_{ii(1)} I_{i(1)} + \cdots + Z_{in(1)} I_{n(1)} \right]$$

[1 at suffix represents the positive sequence quantity]
Because currents at all buses except the i-th bus are zero,

$$I_{i(1)} \neq 0; \ I_{1(1)} = I_{2(1)} = \cdots = I_{k(1)} = \cdots = I_{n(1)} = 0$$

$$\therefore V_{i(1)} = E - \left[0 + \cdots + Z_{ii(1)} I_{i(1)} + \cdots + 0 \right]$$

$$\text{i.e., } \ I_{i(1)} = \frac{E}{Z_{ii(1)}} \tag{2.64}$$

Assuming the series fault impedance to be Z_f,

$$I_{i(1)} = \frac{E}{Z_f + Z_{ii(1)}} \tag{2.65}$$

The bus voltage at i-th bus would be zero due to direct fault at i-th bus. The bus voltage at other bus are given by,

$$V_{k(1)} = E - Z_{ki(1)} I_{i(1)}$$

$$= E - Z_{ki(1)} \frac{E}{Z_f + Z_{ii(1)}}$$

$$= E \left[1 - \frac{Z_{ki(1)}}{Z_f + Z_{ii(1)}} \right]$$

$$\therefore V_{k(1)} = E \left[\frac{Z_f + Z_{ii(1)} - Z_{ki(1)}}{Z_f + Z_{ii(1)}} \right] \tag{2.66}$$

2.13.2 Single Line to Ground Fault

Let single phase to ground fault (phase "a" is shorted with ground through Z_f, the fault impedance) be occurring at bus i (Figure 2.23).

$$\text{Here,} \, I_{ib} = I_{ic} = 0 \tag{2.67}$$

$$V_{ia} = I_{ia}Z_f \tag{2.68}$$

Because $I_{ib} = I_{ic}$, following Equation (2.25a),

$$
\begin{aligned}
&I_{i(0)} + a^2 I_{i(1)} + a I_{i(2)} = I_{i(0)} + a I_{i(1)} + a^2 I_{i(2)} \\
&\text{or,} \ \left(a^2 - a\right) I_{i(1)} = \left(a^2 - a\right) I_{i(2)} \\
&\therefore I_{i(1)} = I_{i(2)}
\end{aligned}
\tag{2.69}
$$

Because $I_{ib} = 0$,

$$
\begin{aligned}
&I_{i(0)} + a^2 I_{i(1)} + a I_{i(2)} = 0 \\
&\text{i.e.,} \ I_{i(0)} = -\left(a^2 + a\right) I_{i(1)} \quad \left[\because I_{i(1)} = I_{i(2)}\right] \\
&\text{or,} \ I_{i(0)} = I_{i(1)} \quad \left[\because \left(1 + a + a^2\right) = 0\right]
\end{aligned}
\tag{2.70}
$$

With reference to Figure 2.23,

$$
\begin{aligned}
&V_{ia} = I_{ia}Z_f \\
&V_{i(0)} + V_{i(1)} + V_{i(2)} = Z_f\left[I_{i(0)} + I_{i(1)} + I_{i(2)}\right] \\
&\text{i.e.,} \ V_{i(0)} + V_{i(1)} + V_{i(2)} = 3I_{i(1)}Z_f
\end{aligned}
\tag{2.71}
$$

$$\left[\because I_{i(0)} = I_{i(1)} = I_{i(2)}, \text{just shown above in Equtions}\left(2.69 \text{ and } 2.70\right)\right.$$

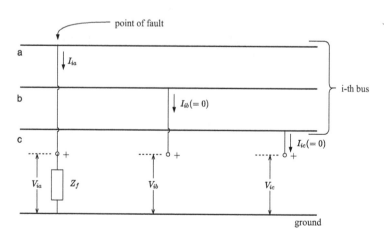

FIGURE 2.23 Line-to-ground fault at i-th bus.

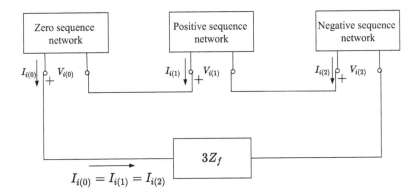

FIGURE 2.24 Interconnection of sequence networks in single line to ground fault.

The equivalent circuit of the Equation (2.71) can then be the network having the series-connected sequence networks across the fault impedance Z_f (Figure 2.24).

In the generalised n – bus network, for a single line-to-ground fault at the i th bus, Equation (2.71) can be rewritten as

$$V_{i(0)} + V_{i(1)} + V_{i(2)} = 3Z_f I_{i(1)} \tag{2.72}$$

$$\text{Also, } I_{k(0)} = I_{k(1)} = I_{k(2)} = 0 \quad \left[\text{for } k = 1, 2, \ldots, n; k \neq i \right]$$

From Equation (2.72) in terms of sequence network quantities we have

$$-Z_{ii(0)} I_{i(0)} + E - Z_{ii(1)} I_{i(1)} - Z_{ii(2)} I_{i(2)} = 3Z_f I_{i(1)}$$

$$\text{or, } \quad E - Z_{ii(0)} I_{i(0)} - Z_{ii(1)} I_{i(1)} - Z_{ii(2)} I_{i(2)} = 3Z_f I_{i(1)} \tag{2.73}$$

$$\text{Since } \quad I_{i(0)} = I_{i(1)} = I_{i(2)}, \text{Equation } (2.73) \text{ reduces to}$$

$$E = I_{i(1)} \left[Z_{ii(0)} + Z_{ii(1)} + Z_{ii(2)} + 3Z_f \right]$$

$$\therefore I_{i(1)} = \frac{E}{Z_{ii(0)} + Z_{ii(1)} + Z_{ii(2)} + 3Z_f} \tag{2.74}$$

The kt^h bus voltage sequence components are given by

$$V_{k(0)} = -Z_{ki(0)} I_{i(0)} = -Z_{ki(0)} I_{i(1)}$$

$$\therefore V_{k(0)} = \frac{-Z_{ki(0)} \cdot E}{Z_{ii(0)} + Z_{ii(1)} + Z_{ii(2)} + 3Z_f} \tag{2.75}$$

$$\text{Also, } V_{k(1)} = E \left[1 - \frac{-Z_{ki(1)}}{Z_{ii(0)} + Z_{ii(1)} + Z_{ii(2)} + 3Z_f} \right]$$

$$\left[\because V_{k(1)} = E - Z_{ki(1)}I_{i(1)}\right]$$

$$\text{or,} \quad V_{k(1)} = \frac{Z_{ii(0)} + Z_{ii(1)} + Z_{ii(2)} + 3Z_f - Z_{ki(1)}}{Z_{ii(0)} + Z_{ii(1)} + Z_{ii(2)} + 3Z_f} \cdot E \tag{2.76}$$

$$\text{and} \quad V_{k(2)} = -Z_{ki(2)}I_{i(2)} = -Z_{ki(2)}I_{i(1)}$$

$$\therefore V_{k(2)} = \frac{-Z_{ki(2)} \cdot E}{Z_{ii(0)} + Z_{ii(1)} + Z_{ii(2)} + 3Z_f} \tag{2.77}$$

$Z_{ki(1)}, Z_{ki(0)}, Z_{ki(2)}, Z_{ii(1)}, Z_{ii(0)},$ and, $Z_{ii(2)}$ are the respective entries of the $[Z_{\text{Bus}}]$ matrix of the n-bus power network in sequence terms. Thus, it is possible to determine the sequence components of fault currents and the sequence components of voltages at any bus of the n-bus power network subjected to a single line-to-ground fault at bus i.

2.13.3   Line-to-Line Fault

Let there be a line to line fault in a three-phase balanced system (Figure 2.25).
Obviously, $I_{ib} = -I_{ic}$

$$\text{i.e.,} \quad I_{i(0)} + a^2 I_{i(1)} + a I_{i(2)} = -\left(I_{i(0)} + a I_{i(1)} + a^2 I_{i(2)}\right)$$

$$\text{or} \quad 2I_{i(0)} + \left(a^2 + a\right)\left(I_{i(1)} + I_{i(2)}\right) = 0 \tag{2.78}$$

$$\text{Also,} \quad V_{ib} = V_{ic} + Z_f I_{ib} \tag{2.79}$$

$$\text{and } I_{ia} = 0$$

$$\text{i.e. } I_{i(0)} + I_{i(1)} + I_{i(2)} = 0 \tag{2.80}$$

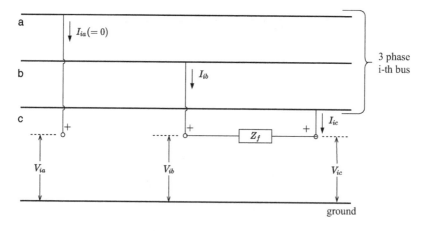

FIGURE 2.25   Double-line fault at i-th bus.

Using Equation (2.80) in (2.78), we have

$$2I_{a(0)} + (-1)(-I_{a(0)}) = 0 \quad \left[\because (a^2 + a) = -1\right]$$

i.e., $3I_{a(0)} = 0$

or, $I_{a(0)} = 0$ \hfill (2.81)

Using Equation (2.81) in (2.80),

$$I_{i(1)} = -I_{i(2)} \hfill (2.82)$$

Because $V_{ib} = V_{ic} + Z_f I_{ib}$, we can write,

$$V_{i(0)} + V_{i(1)} + V_{i(2)} = \left(V_{i(0)} + aV_{i(1)} + a^2 V_{i(2)}\right) + Z_f \left(I_{i(0)} + a^2 I_{i(1)} + aI_{i(2)}\right)$$

Simplification yields,

$$\left(a^2 - a\right) V_{i(1)} = \left(a^2 - a\right) I_{i(1)} Z_f + \left(a^2 - a\right) V_{i(2)}$$

i.e., $V_{i(1)} = I_{i(1)} Z_f + V_{i(2)}$ \hfill (2.83)

The equivalent sequence network circuit is shown in Figure 2.26.

Next, the fault (L-L) is assumed to occur at the i-th bus of the generalised n-bus network. Here $V_{i(0)} = 0$ as zero-sequence network is a passive network.

Using Equation (2.83) for the multi-bus network, we have

$$E_i - Z_{ii(1)} I_{i(1)} = I_{i(1)} Z_f - Z_{ii(2)} I_{i(2)} \hfill (2.84)$$

Since $V_{i(1)} = E_i - Z_{ii(1)} I_{i(1)}$

and $V_{i(2)} = -Z_{ii(2)} I_{i(2)}$

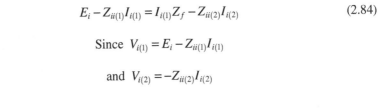

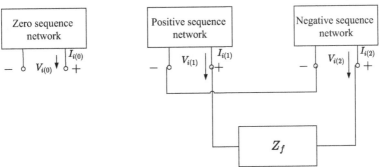

FIGURE 2.26 Interconnection of sequence network for double line (L-L) fault.

Using the relation $I_{i(1)} = -I_{i(2)}$ from Equation (2.84), we obtain,

$$I_{i(1)} = \frac{E_i}{Z_{ii(1)} + Z_{ii(2)} + Z_f} \qquad (2.85)$$

Also,

$$V_{k(1)} = E_k - Z_{ki(1)}I_{i(1)}$$

$$= E_k - Z_{ki(1)} \frac{E_i}{Z_{ii(1)} + Z_{ii(2)} + Z_f}$$

$$= E\left[1 - \frac{Z_{ki(1)}}{Z_{ii(1)} + Z_{ii(2)} + Z_f}\right] \quad \left(\text{assuming } E_i = E_k = E\right)$$

$$\therefore V_{k(1)} = E\left[\frac{Z_{ii(1)} + Z_{ii(2)} + Z_f - Z_{ki(1)}}{Z_{ii(1)} + Z_{ii(2)} + Z_f}\right] \qquad (2.86)$$

and, $V_{k(2)} = -Z_{ki(2)}I_{i(2)} = -Z_{ki(2)}\left(+I_{i(2)}\right) = -Z_{ki(2)}\left(-I_{i(1)}\right) \quad \left[\because I_{i(1)} = -I_{i(2)}\right]$

$$\text{or, } V_{k(2)} = Z_{ki(2)}I_{i(1)} = Z_{ki(2)} \frac{E_k}{Z_{ii(1)} + Z_{ii(2)} + Z_f}$$

$$\text{i.e., } V_{k(2)} = \frac{Z_{ki(2)} \cdot E}{Z_{ii(1)} + Z_{ii(2)} + Z_f}; \qquad [E_k = E] \qquad (2.87)$$

Equation (2.85) represents the i-th bus positive sequence current while the k-th bus sequence voltages are represented in Equations (2.86) and (2.87) in the multi-bus power network subjected to a line-to-line fault at i-th bus.

2.13.4 DOUBLE LINE-TO-GROUND (L-L-G) FAULT

Let a double line-to-ground fault be occurring at bus i as represented in Figure 2.27.

$$\text{Obviously, } I_{ia} = 0 \qquad (2.88)$$

$$V_{ib} = V_{ic} \qquad (2.89)$$

$$\text{and } V_{ib} = \left(I_{ib} + I_{ic}\right)Z_f \qquad (2.90)$$

As $I_{ia} = 0$ (as per Equation (2.88)),

$$I_{i(0)} + I_{i(1)} + I_{i(2)} = 0 \qquad (2.91)$$

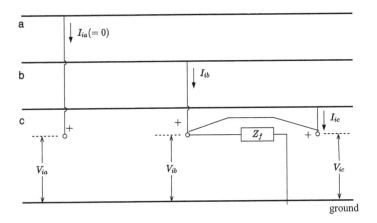

FIGURE 2.27 Double line to ground fault at i-th bus.

From Equation (2.89),

$$V_{i(0)} + a^2 V_{i(1)} + a V_{i(2)} = V_{i(0)} + a V_{i(1)} + a^2 V_{i(2)}$$

Simplification yields,

$$V_{i(1)} = V_{i(2)} \qquad\qquad (2.92)$$

Also from Equation (2.90)

$$V_{i(0)} + a^2 V_{i(1)} + a V_{i(2)} = \left(I_{i(0)} + a^2 I_{i(1)} + a I_{i(2)} + I_{i(0)} + a I_{i(1)} + a^2 I_{i(2)} \right) Z_f$$

Simplification yields

$$V_{i(0)} - V_{i(1)} = \left[2 I_{i(0)} + \left(a^2 + a \right) \left(I_{i(1)} + I_{i(2)} \right) \right] Z_f$$

$$\text{or,} \quad V_{i(0)} - V_{i(1)} = 3 Z_f I_{i(0)}$$

$$V_{i(0)} = V_{i(1)} + 3 Z_f I_{i(0)} \qquad\qquad (2.93)$$

The sequence network interconnection is shown in Figure 2.28.

For the multi-bus network, in case of double line-to-ground fault, the positive sequence network gives

$$V_{i(1)} = E - Z_{ii(1)} I_{i(1)}$$

$$\text{i.e.,} \quad I_{i(1)} = \frac{-V_{i(1)} + E}{Z_{ii(1)}} \qquad\qquad (2.94)$$

$$V_{i(2)} = -Z_{ii(2)} I_{i(2)}$$

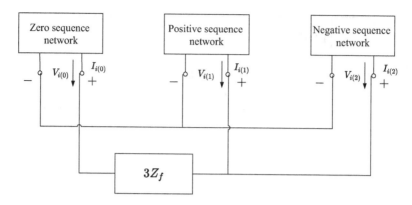

FIGURE 2.28   Sequence network interconnection for double line to ground fault.

$$\text{i.e.,} \quad I_{i(2)} = \frac{-V_{i(2)}}{Z_{ii(2)}} = \frac{-V_{i(1)}}{Z_{ii(2)}} \quad \left[\text{using Equation (2.92)}\right] \qquad (2.95)$$

From Equation (2.93), we can write

$$V_{i(1)} = -Z_{ii(0)}I_{i(0)} - 3Z_f I_{i(0)} \quad \left[\because V_{i(0)} = -Z_{ii}I_{i(0)}\right]$$

$$\text{i.e.,} \quad I_{i(0)} = -\frac{V_{i(1)}}{Z_{ii(0)} + 3Z_f} \qquad (2.96)$$

Substituting the values of $I_{i(1)}$, $I_{i(2)}$ and $I_{i(0)}$ from Equations (2.94)–(2.96), respectively, in Equation (2.91), we get

$$-\frac{V_{i(1)}}{Z_{ii(0)} + 3Z_f} + \frac{E - V_{i(1)}}{Z_{ii(1)}} - \frac{V_{i(1)}}{Z_{ii(2)}} = 0$$

$$\text{or,} \quad \frac{V_{i(1)}}{Z_{ii(0)} + 3Z_f} + \frac{V_{i(1)} - E}{Z_{ii(1)}} + \frac{V_{i(1)}}{Z_{ii(2)}} = 0$$

$$\therefore V_{i(1)} = \frac{Z_{ii(2)}\left(Z_{ii(0)} + 3Z_f\right)E}{Z_{ii(1)}Z_{ii(2)} + Z_{ii(1)}\left(Z_{ii(0)} + 3Z_f\right) + Z_{ii(2)}\left(Z_{ii(0)} + 3Z_f\right)}$$

$$\text{i.e.,} \quad V_{i(1)} = \frac{Z_{ii(2)}\left[Z_{ii(0)} + 3Z_f\right]}{\Delta Z} \qquad (2.97)$$

where, $\Delta Z = Z_{ii(1)}Z_{ii(2)} + Z_{ii(1)}\left(Z_{ii(0)} + 3Z_f\right) + Z_{ii(2)}\left(Z_{ii(0)} + 3Z_f\right)$

Substituting the value of $V_{i(1)}$ from Equation (2.97) in Equations (2.94), (2.95) and (2.96), the expressions of $I_{i(0)}$, $I_{i(1)}$ and $I_{i(2)}$ are modified as follows:

$$I_{i(0)} = -\frac{V_{i(1)}}{Z_{ii(0)} + 3Z_f} = -\frac{Z_{ii(2)} \cdot E}{\Delta Z} \tag{2.98}$$

$$I_{i(1)} = \frac{E - V_{i(1)}}{Z_{ii(1)}} = \frac{E - \left[\dfrac{Z_{ii(2)}\left(Z_{ii(0)} + 3Z_f\right)E}{\Delta Z}\right]}{Z_{ii(1)}} = \frac{E\left(Z_{ii(2)} + Z_{ii(0)} + 3Z_f\right)}{\Delta Z} \tag{2.99}$$

$$\text{and,} \quad I_{i(2)} = \frac{-V_{i(1)}}{Z_{ii(2)}} = \frac{-\left[Z_{ii(0)} + 3Z_f\right]}{\Delta Z} \tag{2.100}$$

The sequence voltages at bus k are obtained as

$$V_{k(0)} = -Z_{ki(0)}I_{i(0)} = \frac{Z_{ki(0)}\,Z_{ii(2)}E}{\Delta Z} \tag{2.101}$$

Here, $E = E_i = E_k$

$$V_{k(1)} = E - Z_{ki(1)}I_{i(1)} = \frac{\Delta Z - Z_{ki(1)}\left(Z_{ii(2)} + 3Z_f + Z_{ii(0)}\right)E}{\Delta Z} \tag{2.102}$$

$$\text{and} \quad V_{k(2)} = -Z_{ki(2)}I_{i(2)}$$

$$\text{or,} \quad V_{k(2)} = \frac{Z_{ki(2)}\left(3Z_f + Z_{ii(0)}\right)E}{\Delta Z} \tag{2.103}$$

2.13.5 SEQUENCE CURRENTS IN THE INTERCONNECTING LINE BETWEEN THE FAULTED BUS I AND HEALTHY BUS K

The sequence currents in the interconnecting line between faulted bus i and healthy bus k are given as

$$I_{ik(0)} = \frac{V_{i(0)} - V_{k(0)}}{Z_{ik(0)}}; \quad I_{ik(1)} = \frac{V_{i(1)} - V_{k(1)}}{Z_{ik(1)}}; \quad \text{and} \quad I_{ik(2)} = \frac{V_{i(2)} - V_{k(2)}}{Z_{ik(2)}} \tag{2.104}$$

$$\text{Also,} \quad I_{ik(0)} = -I_{ik(0)}$$

$$I_{ik(1)} = -I_{ik(1)}$$

$$\text{and} \quad I_{ik(2)} = -I_{ik(2)}$$

$-I_{ik(0)}$, $-I_{ik(1)}$ and $-I_{ik(2)}$ being the sequence currents flowing from the k-th bus to the i-th bus and are usually negative of $I_{ik(0)}$, $I_{ik(1)}$ and, $I_{ik(2)}$, respectively, except the condition when there is a star-delta transformer in the line.

Example 2.9

For a three-bus power system, the sequence impedances are given as follows:

$$[Z_{Bus}]_{(1)} = [Z_{Bus}]_{(2)} = \begin{bmatrix} j0.18 & j0.12 & j0.15 \\ j0.12 & j0.18 & j0.15 \\ j0.15 & j0.15 & j0.25 \end{bmatrix} \text{p.u.}$$

$$[Z_{Bus}]_{(0)} = \begin{bmatrix} j0.09 & j0.06 & j0.08 \\ j0.06 & j0.09 & j0.08 \\ j0.08 & j0.08 & j0.14 \end{bmatrix} \text{p.u.}$$

Assuming the pre-fault bus voltage to be 1.00 p.u./phase, compute the fault current, bus voltages and line currents in the post-fault condition for (i) single line-to-ground solid fault, (ii) line-to-line short circuit fault, (iii) solid three-phase-to-ground fault and (iv) solid double line-to-ground fault, assuming the faults occurring separately at bus number 3.

Line no.	From Bus	To Bus	Zero-Sequence Line Impedance	Positive-Sequence Line Impedance	Negative-Sequence Line Impedance
1	1	2	$j0.15$	$j0.3$	$j0.3$
2	2	3	$j0.15$	$j0.3$	$j0.3$
3	1	3	$j0.15$	$j0.3$	$j0.3$

Solution

I) FOR SINGLE LINE-TO-GROUND FAULT

Assuming the fault occurring in bus number 3,

$$I_{3(0)} = I_{3(1)} = I_{3(2)} = \frac{E}{Z_{33(0)} + Z_{33(1)} + Z_{33(2)}} \quad \left[\because Z_f = 0, \text{ for solid fault} \right]$$

Here, $Z_{33(1)} = Z_{33(2)}$, as given in the statement of this problem. As $E = 1$ p.u. we have,

$$I_{3(0)} = I_{3(1)} = I_{3(2)} = \frac{1}{Z_{33(0)} + 2Z_{33(1)}} = \frac{1}{j0.14 + 2 \times j0.25} = -j1.56 \text{ p.u.}$$

$$\therefore I_{3(\text{fault})} = 3I_{a1} = -j\,3 \times 1.56 = -j4.68 \text{ p.u.}$$

Next, we determine the bus voltages after the fault occurs.

$$V_{1(0)} = -Z_{13(0)}I_{3(0)} = -j0.08(-j1.56) = -0.125 \text{ p.u.}$$

$$V_{2(0)} = -Z_{23(0)}I_{3(0)} = -j0.08(-j1.56) = -0.125 \text{ p.u.}$$

$$V_{3(0)} = -Z_{33(0)}I_{3(0)} = -j0.14(-j1.56) = -0.218 \text{ p.u.} \approx -0.22 \text{ p.u.}$$

$$V_{1(1)} = E - Z_{13(1)}I_{3(1)} = 1 - j0.15(-j1.56) = 0.766 \text{ p.u.}$$

$$V_{1(2)} = -Z_{13(2)}I_{3(2)} = -j0.15(-j1.56) = -0.234 \text{ p.u.}$$

$$V_{2(1)} = E - Z_{23(1)}I_{3(1)} = 1 - j0.15(-j1.56) = 0.766 \text{ p.u.}$$

$$V_{3(1)} = E - Z_{33(1)}I_{3(1)} = 1 - j0.25(-j1.56) = 0.61 \text{ p.u.}$$

$$V_{1(2)} = -Z_{13(2)}I_{3(2)} = -j0.15(-j1.56) = -0.234 \text{ p.u.}$$

$$V_{2(2)} = -Z_{23(2)}I_{3(2)} = -j0.15(-j1.56) = -0.234 \text{ p.u.}$$

$$V_{3(2)} = -Z_{33(2)}I_{3(2)} = -j0.25(-j1.56) = -0.39 \text{ p.u}$$

Next, we compute line currents (here, suffix L stands for the p.u. impedances of respective lines)

$$I_{12(0)} = \frac{V_{1(0)} - V_{2(0)}}{Z_{12(0)L}} = 0 \quad \left[\because V_{1(0)} = V_{2(0)} \right]$$

$$I_{23(0)} = \frac{V_{2(0)} - V_{3(0)}}{Z_{23(0)L}} = \frac{-0.125 - (-0.218)}{j0.15} = -j0.62 \text{ p.u.}$$

$$I_{31(0)} = \frac{V_{3(0)} - V_{1(0)}}{Z_{31(0)L}} = \frac{-0.218 - (-0.125)}{j0.15} = j0.62 \text{ p.u.}$$

$$I_{12(1)} = \frac{V_{1(1)} - V_{2(1)}}{Z_{12(1)L}} = 0 \quad \left[\because V_{1(1)} = V_{2(1)} \right]$$

$$I_{23(1)} = \frac{V_{2(1)} - V_{3(1)}}{Z_{23(1)L}} = \frac{0.766 - 0.61}{j0.3} = -j0.52 \text{ p.u.}$$

$$I_{31(1)} = \frac{V_{3(1)} - V_{1(1)}}{Z_{13(1)L}} = \frac{0.61 - 0.766}{j0.3} = j0.52 \text{ p.u.}$$

$$I_{12(2)} = \frac{V_{1(2)} - V_{2(2)}}{Z_{12(2)L}} = 0 \quad \left[\because V_{1(2)} = V_{2(2)} \right]$$

$$I_{23(2)} = \frac{V_{2(2)} - V_{3(2)}}{Z_{23(2)L}} = \frac{0.234 - (-0.39)}{j0.3} = -j0.52 \text{ p.u.}$$

$$I_{31(2)} = \frac{V_{3(2)} - V_{1(2)}}{Z_{31(2)L}} = \frac{-0.39 - (-0.234)}{j0.3} = j0.52 \text{ p.u.}$$

$\therefore$ Finally, for L-G fault at bus 3, we get the following sequence quantities.

$$V_{1(0)} = -0.125 \text{ p.u.}; \quad V_{1(1)} = 0.766 \text{ p.u.}; \quad V_{1(2)} = -0.234 \text{ p.u.}$$

$$V_{2(0)} = -0.125 \text{ p.u.}; \quad V_{2(1)} = 0.766 \text{ p.u.}; \quad V_{2(2)} = -0.234 \text{ p.u.}$$

$$V_{3(0)} = -0.218 \text{ p.u.}; \quad V_{3(1)} = 0.61 \text{ p.u.}; \quad V_{3(2)} = -0.39 \text{ p.u.}$$

$$I_{3(0)} = I_{3(1)} = I_{3(2)} = -j1.56 \text{ p.u.}; \quad \left[I_{3(\text{fault})} = -j4.68 \text{ p.u.} \right]$$

$$I_{12(0)} = 0 \text{ p.u.}; \quad I_{12(1)} = 0 \text{ p.u.}; \quad I_{12(2)} = 0 \text{ p.u.}$$

$$I_{23(0)} = -j0.62 \text{ p.u.}; \quad I_{23(1)} = -j0.52 \text{ p.u.}; \quad I_{23(2)} = -j0.52 \text{ p.u.}$$

$$I_{31(0)} = -j0.62 \text{ p.u.}; \quad I_{31(1)} = j0.52 \text{ p.u.}; \quad I_{31(2)} = j0.52 \text{ p.u.}$$

II) FOR LINE-TO-LINE FAULT

For the line-to-line fault at bus 3,

$$I_{3(1)} = \frac{E}{Z_{33(1)} + Z_{33(2)} + Z_f}$$

Here, $Z_{33(1)} = Z_{33(2)} = j0.25 \text{ p.u.}$; $Z_f = 0$ and $E = 1 \text{ p.u.}$

$$\therefore I_{3(1)} = \frac{1}{j0.25 + j0.25} = \frac{1}{j0.5} = -j2 \text{ p.u.}$$

$$I_{3(2)} = I_{3(1)} = j2 \text{ p.u.}$$

The zero-sequence component is absent in line-to-line fault.

$$\therefore I_{3(0)} = 0$$

Post-fault bus voltages are given as:

$$V_{1(1)} = E - Z_{13(1)}I_{3(1)} = 1 - j0.15(-j2) = 0.7 \text{ p.u.}$$

$$V_{1(2)} = -Z_{13(2)}I_{3(2)} = -j0.15 \times j2 = 0.3 \text{ p.u.}$$

$$V_{1(0)} = 0 \text{ p.u.}$$

$$V_{2(1)} = E - Z_{23(1)}I_{3(1)} = 1 - j0.15(-j2) = 0.7 \text{ p.u.}$$

$$V_{2(2)} = -Z_{23(2)}I_{3(2)} = -j0.15 \times j2 = 0.3 \text{ p.u.}$$

$$V_{2(0)} = 0 \text{ p.u.}$$

$$V_{3(1)} = E - Z_{33(1)}I_{3(1)} = 1 - (j0.25)(-j2) = 0.5 \text{ p.u.}$$

$$V_{3(2)} = -Z_{33(2)}I_{3(2)} = -j0.25 \times j2 = 0.5 \text{ p.u.}$$

$$V_{3(0)} = 0$$

Line currents are obtained as (suffix L stands for respective line impedances)

$$I_{23(1)} = \frac{V_{2(1)} - V_{3(1)}}{Z_{23(1)L}} = \frac{0.7 - 0.5}{j0.3} = -j0.67 \text{ p.u.}$$

$$I_{23(2)} = \frac{V_{2(2)} - V_{3(2)}}{Z_{23(2)L}} = \frac{0.3 - 0.5}{j0.3} = -j0.67 \text{ p.u.}$$

$$I_{23(0)} = 0 \text{ p.u.}$$

$$I_{13(1)} = \frac{V_{1(1)} - V_{3(1)}}{Z_{13(1)L}} = \frac{0.7 - 0.5}{j0.3} = -j0.67 \text{ p.u.} = -I_{31(1)}$$

$$I_{13(2)} = \frac{V_{1(2)} - V_{3(2)}}{Z_{13(2)L}} = \frac{0.3 - 0.5}{j0.3} = j0.67 \text{ p.u.} = -I_{31(2)}$$

$$I_{13(0)} = 0 \text{ p.u.} = -I_{31(0)}$$

$$I_{12(1)} = \frac{V_{1(1)} - V_{2(1)}}{Z_{12(1)L}} = \frac{0.7 - 0.7}{j0.3} = 0 \text{ p.u.}$$

$$I_{12(2)} = \frac{V_{1(2)} - V_{2(2)}}{Z_{12(2)L}} = \frac{0.3 - 0.3}{j0.3} = 0 \text{ p.u.}$$

$$I_{12(0)} = 0$$

∴ For line-to-line fault we have

$$V_{1(0)} = 0 \text{ p.u.}; \quad V_{1(1)} = 0.7 \text{ p.u.}; \quad V_{1(2)} = 0.3 \text{ p.u.}$$

$$V_{2(0)} = 0 \text{ p.u.}; \quad V_{2(1)} = 0.7 \text{ p.u.}; V_{2(2)} = 0.3 \text{ p.u.}$$

$$V_{3(0)} = 0 \text{ p.u.}; \quad V_{3(1)} = 0.5 \text{ p.u.}; \quad V_{3(2)} = 0.5 \text{ p.u.}$$

$$I_{3(1)} = -j2 \text{ p.u.}; I_{3(2)} = j2 \text{ p.u.}; I_{3(0)} = 0$$

$$I_{12(0)} = 0 \text{ p.u.}; I_{12(1)} = 0 \text{ p.u.}; I_{12(2)} = 0 \text{ p.u.}$$

$$I_{23(0)} = 0 \text{ p.u.}; I_{23(1)} = -j0.67 \text{ p.u.}; \quad I_{23(2)} = j0.67 \text{ p.u.}$$

$$I_{31(0)} = 0; \quad I_{31(1)} = j0.67 \text{ p.u.}; I_{31(2)} = -j0.67 \text{ p.u.}$$

III) FOR SOLID THREE-PHASE-TO-GROUND FAULT

The fault current is given by

$$I_{3(1)} = \frac{E}{Z_{33(1)}} \qquad [\because Z_f = 0]$$

$$\text{or, } I_{3(1)} = \frac{1}{j0.25} = -j4 \text{ p.u.}$$

Post-fault bus voltages are given by

$$V_{1(1)} = E - Z_{13(1)}I_{3(1)} = 1 - j0.15(-j4) = 0.4 \text{ p.u.}$$

$$V_{2(1)} = E - Z_{23(1)}I_{3(1)} = 1 - j0.15(-j4) = 0.4 \text{ p.u.}$$

$$V_{3(1)} = E - Z_{33(1)}I_{3(1)} = 1 - j0.25(-j4) = 0 \text{ p.u.}$$

Line currents during the fault period are given by

$$I_{12(1)} = \frac{V_{1(1)} - V_{2(1)}}{Z_{12(1)L}} = 0 \quad \left[\because V_{1(1)} = V_{2(1)}\right]$$

$$I_{23(1)} = \frac{V_{2(1)} - V_{3(1)}}{Z_{23(1)L}} = \frac{0.4 - 0}{j0.3} = -j1.33 \text{ p.u.}$$

$$I_{13(1)} = \frac{V_{1(1)} - V_{3(1)}}{Z_{13(1)L}} = \frac{0.4 - 0}{j0.3} = -j1.33 \text{ p.u.} = -I_{31(1)}$$

Finally, we have for a three-phase-to-ground fault, $I_{3(1)} = -j4$ p.u. [only positive sequence current flows in balanced three-phase faulted network]

$$V_{1(0)} = 0 \text{ p.u.; } V_{1(1)} = 0.4 \text{ p.u.; } V_{1(2)} = 0 \text{ p.u.}$$

$$V_{2(0)} = 0 \text{ p.u.; } V_{2(1)} = 0.4 \text{ p.u.; } V_{2(2)} = 0 \text{ p.u.}$$

$$V_{3(0)} = 0 \text{ p.u.; } V_{3(1)} = 0 \text{ p.u.; } V_{3(2)} = 0 \text{ p.u.}$$

Also, $I_{12(1)} = 0$ p.u.; $I_{23(1)} = -j1.33$ p.u. ; $I_{31(1)} = j1.33$ p.u.

IV) FOR DOUBLE-LINE-TO-GROUND FAULT

The fault current is given by

$$I_{31(1)} = \frac{E\left(Z_{33(2)} - Z_{33(0)}\right)}{\Delta Z}, \qquad \text{using equation (2.99) with } i = 3 \text{ and } Z_f = 0$$

$$= \frac{1(j0.25 + j0.14)}{j0.25 \times j0.25 + j0.25 \times j0.14 + j0.25 \times j0.14}$$

$[\because \Delta Z = Z_{33(1)}Z_{33(2)} + Z_{33(1)}Z_{33(0)} + Z_{33(2)}Z_{33(0)} \quad \text{for } Z_f = 0 \text{ and } i = 3 \text{ from Equation}$ (2.97)]

$$\therefore I_{31(1)} = \frac{j0.39}{-0.1325} = -j2.94 \text{ p.u.}$$

Also, $I_{3(2)} = -\dfrac{\left(Z_{33(0)}\right)}{\Delta Z}$, with $i = 3$ and $Z_f = 0$ in Equation (2.100)

$$\text{i.e.,}\quad I_{3(2)} = -\frac{j0.14}{-0.1325} = j1.06 \text{ p.u.}$$

and, $I_{3(0)} = -\dfrac{\left(Z_{33(2)}\right)E}{\Delta Z}$, with $i = 3$ and $Z_f = 0$ in Equation (2.98)

$$= -\frac{j0.25 \times 1}{-0.1325} = j1.89 \text{ p.u.}$$

Bus voltages after the occurrence of the fault can be calculated as follows.

$$V_{1(1)} = E - Z_{13(1)}I_{3(1)} = 1 - j0.15(-j2.94) = 0.559 \text{ p.u.}$$

$$V_{1(2)} = -Z_{13(2)}I_{3(2)} = -0.15(j1.06) = 0.159 \text{ p.u.}$$

$$V_{1(0)} = -Z_{13(0)}I_{3(0)} = -j0.08(j1.89) = 0.1512 \text{ p.u.}$$

$$V_{2(1)} = E - Z_{23(1)}I_{3(1)} = 1 - j0.15(-j2.94) = 0.559 \text{ p.u.}$$

$$V_{2(2)} = -Z_{23(2)}I_{3(2)} = -j0.15(j1.06) = 0.159 \text{ p.u.}$$

$$V_{2(0)} = -Z_{23(0)}I_{3(0)} = -j0.08(j1.89) = 0.1512 \text{ p.u.}$$

$$V_{3(1)} = E - Z_{33(1)}I_{3(1)} = 1 - j0.25(-j2.94) = 0.265 \text{ p.u.}$$

$$V_{3(2)} = -Z_{33(2)}I_{3(2)} = -j0.25(j1.06) = 0.265 \text{ p.u.}$$

$$V_{3(0)} = -Z_{33(0)}I_{3(0)} = -j0.14(j1.89) = 0.2646 \text{ p.u.}$$

Line currents can be calculated as follows:

$$I_{12(1)} = \frac{V_{1(1)} - V_{2(1)}}{Z_{12(1)L}} = \frac{0.559 - 0.559}{j0.3} = 0 \text{ p.u.}$$

$$I_{12(2)} = \frac{V_{1(2)} - V_{2(2)}}{Z_{12(2)L}} = \frac{0.159 - 0.159}{j0.3} = 0$$

$$I_{12(0)} = \frac{V_{1(0)} - V_{2(0)}}{Z_{12(0)L}} = \frac{0.1512 - 0.1512}{j0.15} = 0$$

$$I_{13(1)} = \frac{V_{1(1)} - V_{3(1)}}{Z_{13(1)L}} = \frac{0.559 - 0.265}{j0.3} = -j0.98 \text{ p.u.} = -I_{31(1)}$$

$$I_{13(2)} = \frac{V_{1(2)} - V_{3(2)}}{Z_{13(2)L}} = \frac{0.159 - 0.265}{j0.3} = j0.353 \text{ p.u.} = -I_{31(2)}$$

$$I_{13(0)} = \frac{V_{1(0)} - V_{3(0)}}{Z_{13(0)L}} = \frac{0.1512 - 0.2646}{j0.15} = j0.756 \text{ p.u.} = -I_{31(0)}$$

$$I_{23(1)} = \frac{V_{2(1)} - V_{3(1)}}{Z_{23(1)L}} = \frac{0.559 - 0.265}{j0.3} = -j0.98 \text{ p.u.}$$

$$I_{23(2)} = \frac{V_{2(2)} - V_{3(2)}}{Z_{23(2)L}} = \frac{0.159 - 0.265}{j0.3} = j0.353 \text{ p.u.}$$

$$I_{23(0)} = \frac{V_{2(0)} - V_{3(0)}}{Z_{23(0)L}} = \frac{0.1512 - 0.2646}{j0.15} = j0.756 \text{ p.u.}$$

∴ For L-L-G fault at bus 3 we get the following sequence quantities,

$$V_{1(0)} = 0.1512 \text{ p.u.}; \ V_{1(1)} = 0.559 \text{ p.u.}; \ V_{1(2)} = 0.159 \text{ p.u.}$$

$$V_{2(0)} = 0.1512 \text{ p.u.}; \ V_{2(1)} = 0.559 \text{ p.u.}; \ V_{2(2)} = 0.159 \text{ p.u.}$$

$$V_{3(0)} = 0.2646 \text{ p.u.}; \ V_{3(1)} = 0.265 \text{ p.u.}; \ V_{3(2)} = 0.265 \text{ p.u.}$$

$$I_{3(1)} = -j2.94 \text{ p.u.}; \ I_{3(2)} = j1.06 \text{ p.u.}; \ I_{3(0)} = j1.89 \text{ p.u.}$$

$$I_{12(0)} = 0 \text{ p.u.}; \ I_{12(1)} = 0 \text{ p.u.}; \ I_{12(2)} = 0 \text{ p.u.}$$

$$I_{23(0)} = j0.756 \text{ p.u.}; \ I_{23(1)} = -j0.98 \text{ p.u.}; \ I_{23(2)} = j0.353 \text{ p.u.}$$

$$I_{31(0)} = -j0.756; \ I_{31(1)} = j0.98 \text{ p.u.}; \ I_{31(2)} = -j0.353 \text{ p.u.}$$

2.14 SEVERITY OF FAULT CURRENTS AND THE EFFECT OF NEUTRAL GROUNDING REACTANCE

In the three-phase system, let the fault occurs at phase 'a'. The type of fault being L-G fault and Z_f, the fault impedance being zero, the fault current is the given by

$$I_{af} \left(\text{the fault current in phase a} \right) = 3 \left[\frac{E}{Z_{ii(0)} + Z_{ii(1)} + Z_{ii(2)}} \right],$$

As $I_{(0)} = I_{(1)} = I_{(2)} = \frac{1}{3} I_{af}$, for the single line-to-ground fault, $I_{af} = 3 \times$ sequence current.

Assuming the resistance component of impedances to be negligible and the positive sequence reactance being equal to the negative-sequence reactance, we get the modified fault current expression. Here

$$I_{af} = 3 \left[\frac{E}{2X_{ii(1)} + X_{ii(0)}} \right] \tag{2.105}$$

On the other hand, for a solid three-phase fault (where $Z_f \to 0$), we have

$$I_{af} = \frac{E}{X_{ii(1)}} \tag{2.106}$$

Comparing Equations (2.105) and (2.106), we find that for a three-phase system where the neutral is solidly grounded and the positive reactance is equal to the negative reactance and the zero-sequence reactance is of low value, we have the single line-to-ground fault more severe than the three-phase fault. However, if the zero-sequence reactance $\left(X_{ii(0)}\right)$ is of much higher value than the positive-sequence reactance $X_{ii(1)}$, there is a possibility that the three-phase fault is more severe than the single line-to-ground fault.

For an alternator, usually the positive- and negative-sequence reactances are equal while the zero-sequence reactance is of lower magnitude. To make the fault current due to the L-G fault lesser than even the fault current due to three-phase faults, we need to include a grounded reactance (X_n) at the neutral so that

$$\frac{3E}{2X_{ii(1)} + X_{ii(0)} + 3X_n} < \frac{E}{X_{ii(1)}}$$

i.e., $2X_{ii(1)} + X_{ii(0)} + 3X_n > 3X_{ii(1)}$ \qquad (2.107)

or, $X_n > \dfrac{1}{3}\left(X_{ii(1)} - X_{ii(0)}\right)$

Thus, it is possible to select the grounding reactance (X_n) to limit the L-G fault current lower than the L-L-L fault current using a suitable grounding reactance at the neutral of the generator.

2.15 OPEN CONDUCTOR FAULT

Let the three-phase three-wire line be assumed with *open conductor fault* occurring at F (Figure 2.29).

Let the voltage between broken conductors in each phase be $V_{aa'}$, $V_{bb'}$ and $V_{cc'}$ while the line currents are I_a, I_b and I_c (for generalisation of treatments, the currents are assumed even in the broken phases and their respective values will be set to zero successively during the analysis of broken conductor faults).

$$\text{Here, } I = \begin{bmatrix} I_a \\ I_b \\ I_c \end{bmatrix} ; \quad V = \begin{bmatrix} V_{aa'} \\ V_{bb'} \\ V_{cc'} \end{bmatrix}$$

$$\text{and } \begin{bmatrix} I_a \\ I_b \\ I_c \end{bmatrix} = \begin{bmatrix} 1 & 1 & 1 \\ 1 & a^2 & a \\ 1 & a & a^2 \end{bmatrix} \begin{bmatrix} I_{a0} \\ I_{a1} \\ I_{a2} \end{bmatrix}$$

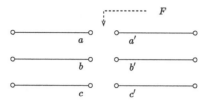

FIGURE 2.29   A three-phase three-wire line with broken conductors.

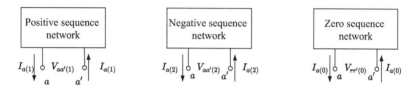

FIGURE 2.30   Sequence network of open conductor fault.

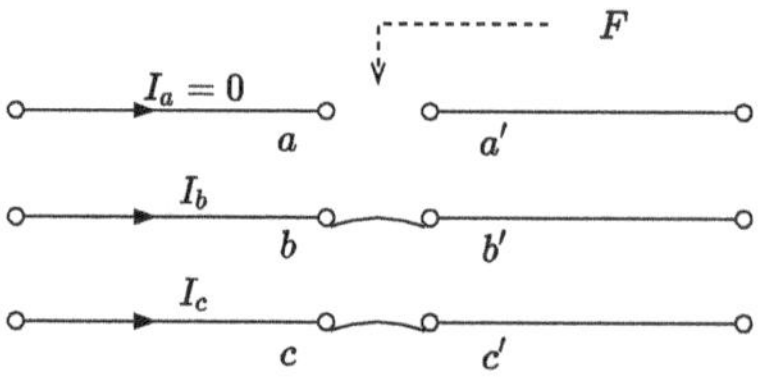

FIGURE 2.31   Single conductor fault at a three-phase three-wire system.

$$\begin{bmatrix} V_a \\ V_b \\ V_c \end{bmatrix} = \begin{bmatrix} 1 & 1 & 1 \\ 1 & a^2 & a \\ 1 & a & a^2 \end{bmatrix} \begin{bmatrix} V_{a0} \\ V_{a1} \\ V_{a2} \end{bmatrix}$$

The sequence component of currents and voltages are shown in Figure 2.30.

2.15.1   SINGLE CONDUCTOR OPEN FAULT

Let us now assume that phase "a" is only open (Figure 2.31).
    Here,

$$V_{bb'} = V_{cc'} = 0 \tag{2.108}$$

$$I_a = 0 \tag{2.109}$$

Using Equations (2.109) and (2.108) in the concept of symmetrical theory related to Equations (2.30) and (2.25b), we get

$$V_{aa'(1)} = V_{aa'(2)} = V_{aa'(0)} = \frac{1}{3} V_{aa'} \tag{2.110}$$

$$\text{and } I_{a(1)} + I_{a(2)} + I_{a(0)} = 0 \tag{2.111}$$

Thus, the sequence network is a partially connected network of individual sequence networks (Figure 2.32).

2.15.2 Two Conductor Open Fault

Let us assume conductor "b" and "c" are open (Figure 2.33).

$$\text{Obviously, } V_{aa'} = 0; \ I_b = I_c = 0$$

From the concept of symmetrical theory, we have,

$$V_{aa'(1)} + V_{aa'(2)} + V_{aa'(0)} = 0 \tag{2.112}$$

$$I_{a(1)} = I_{a(2)} = I_{a(0)} = \frac{1}{3} I_a \tag{2.113}$$

Thus, the sequence network connections would be a series connected network, as shown in Figure 2.34.

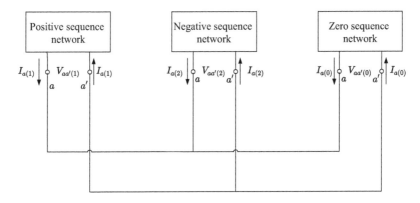

FIGURE 2.32 Connection of sequence network for one conductor open circuit fault.

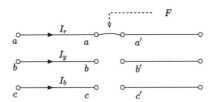

FIGURE 2.33 Open conductor fault at two phases of a three-phase system.

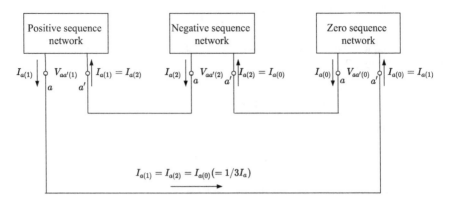

FIGURE 2.34   Interconnection diagram of sequence networks for two conductor open fault.

REFERENCE

1. A. Chakrabarti and S. Halder, *"Power System Analysis: Operation and Control"*, Third Edition, New Delhi: PHI Learning Pvt. Ltd.

3 Contingency Analysis

3.1 INTRODUCTION

Contingency analysis is a method to predict steady state bus voltages and line currents in a power system network following switching (addition or removal) of a line in that network. It helps in checking the components of the system (e.g. bus or line) being subjected to overloading or overvoltage/undervoltage conditions following switching of the prescribed line. Contingency analysis is mostly an *offline* method and usually in the analysis, line resistance, presence of off-nominal tap ratios of transformers and line charging effects are neglected. The linear model of the system is usually assumed where the *principle of superposition* can be applied. The results, thus, have little approximation. Contingency analysis [1] frequently uses *bus impedance matrix* $[Z_{BUS}]$ where loads are to be treated as constant current injections. Addition of a line in the system is simply the addition of an impedance, while removal of the line is accomplished by adding a negative impedance in the prescribed part of the network. The diagonal elements of $[Z_{BUS}]$ provide important characteristics of the power network and facilitate the use of *Thevenin's equivalent impedance* at designated buses. Application of $[Z_{BUS}]$ is evident in power system contingency analysis.

3.2 RELATIONSHIP BETWEEN THEVENIN'S THEOREM AND BUS IMPEDANCE MATRIX $[Z_{BUS}]$

Figure 3.1 shows a power network with buses numbered from 0 to n, 0th bus being the reference bus. Let us assume that bus q is fed by a current source of strength ΔI_q p.u.

Let $V_{10}, V_{20}, \ldots V_{p0}, V_{q0}, \ldots V_{n0}$ be the initial circuit voltage of the respective buses $1, 2, \ldots, p, q, \ldots n$ (the voltage between the respective buses and the reference). In this power network, in general we can write

$$[V_0] = [Z_{BUS}][I_0] \tag{3.1}$$

where $[V_0]$ is the *column vector of initial bus voltages* and $[I_0]$ is the *column vector of initial respective bus currents*. However, due to current injection at bus q, there will be changes in the bus voltages, and the modified governing equation of the system is expressed as

$$[V] = [Z_{BUS}][I_0 + \Delta I] = [Z_{BUS}][I_0] + [Z_{BUS}][\Delta I]$$
$$= [V_0] + [\Delta V] \tag{3.2}$$

$$\text{where,} \quad [\Delta V] = [Z_{BUS}][\Delta I] \tag{3.3}$$

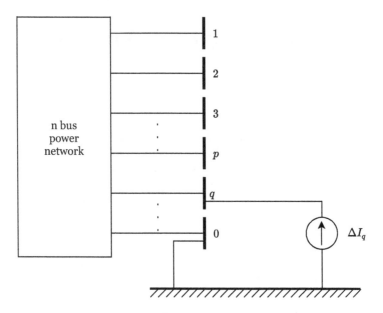

FIGURE 3.1 Current injection at q-th bus of a multi-bus power network.

Because ΔI_q is the only injected current at bus q with no other current injections at any other bus, the column vector $[\Delta I]$ can be expressed as

$$
[\Delta I] = \begin{bmatrix} 0 \\ \vdots \\ 0 \\ \Delta I_q \\ 0 \\ \vdots \\ 0 \end{bmatrix}
\quad \text{while} \quad
[\Delta V] = \begin{bmatrix} \Delta V_1 \\ \Delta V_2 \\ \vdots \\ \Delta V_p \\ \Delta V_q \\ \vdots \\ \Delta V_n \end{bmatrix}
$$

Equation (3.3) can be written as

$$
\begin{bmatrix} \Delta V_1 \\ \Delta V_2 \\ \vdots \\ \Delta V_q \\ \vdots \\ \Delta V_n \end{bmatrix}
=
\begin{bmatrix}
Z_{11} & Z_{12} & \cdots & Z_{1q} & \cdots & Z_{1n} \\
Z_{21} & Z_{22} & \cdots & Z_{2q} & \cdots & Z_{2n} \\
\vdots & \vdots & \cdots & \vdots & \cdots & \vdots \\
Z_{q1} & Z_{q2} & \cdots & Z_{qq} & \cdots & Z_{qn} \\
\vdots & \vdots & \cdots & \vdots & \cdots & \vdots \\
Z_{n1} & Z_{n2} & \cdots & Z_{nq} & \cdots & Z_{nn}
\end{bmatrix}
\begin{bmatrix} 0 \\ 0 \\ \vdots \\ \Delta I_q \\ \vdots \\ 0 \end{bmatrix}
\qquad (3.4)
$$

$$\text{i.e.} \quad \begin{bmatrix} \Delta V_1 \\ \Delta V_2 \\ \vdots \\ \Delta V_q \\ \vdots \\ \Delta V_n \end{bmatrix} = \begin{bmatrix} Z_{1q} \\ Z_{2q} \\ \vdots \\ Z_{qq} \\ \vdots \\ Z_{nq} \end{bmatrix} \Delta I_q \tag{3.5}$$

(as the nonzero quantity in the current matrix exists in the q-th row only).

At the q-th bus, with initial bus voltage V_{q0}, we can express the new q-th bus voltage as

$$V_q = V_{q0} + Z_{qq}\Delta I_q \tag{3.6}$$

Hence, we can interpret Z_{qq} as the Thevenin's impedance at bus q where $Z_{th} \equiv Z_{qq}$. Obviously, Z_{qq} is the *diagonal entry* in the q-th row in $[Z_{BUS}]$ and is the driving point impedance of bus q (Figure 3.2).

The next task is to obtain the Thevenin's impedance between bus p and q so that the bus voltages are $V_1, V_2, \ldots, V_p, V_q, \ldots, V_n$. As $[\Delta V] = [Z_{BUS}][\Delta I]$, we can write

$$\begin{bmatrix} \Delta V_1 \\ \Delta V_2 \\ \vdots \\ \Delta V_p \\ \Delta V_q \\ \vdots \\ \Delta V_n \end{bmatrix} = \begin{bmatrix} Z_{11} & Z_{12} & \cdots & Z_{1p} & Z_{1q} & \cdots & Z_{1n} \\ Z_{21} & Z_{22} & \cdots & Z_{2p} & Z_{2q} & \cdots & Z_{2n} \\ \vdots & \vdots & & \vdots & \vdots & & \vdots \\ Z_{p1} & Z_{p2} & \cdots & Z_{pp} & Z_{pq} & \cdots & Z_{pn} \\ Z_{q1} & Z_{q2} & \cdots & Z_{qp} & Z_{qq} & \cdots & Z_{qn} \\ \vdots & \vdots & & \vdots & \vdots & & \vdots \\ Z_{n1} & Z_{n2} & \cdots & Z_{np} & Z_{nq} & \cdots & Z_{nn} \end{bmatrix} \begin{bmatrix} 0 \\ 0 \\ \vdots \\ \Delta I_p \\ \Delta I_q \\ \vdots \\ 0 \end{bmatrix}$$

$$= \begin{bmatrix} Z_{1p}\Delta I_p + Z_{1q}\Delta I_q \\ Z_{2p}\Delta I_p + Z_{2q}\Delta I_q \\ \vdots \\ Z_{pp}\Delta I_p + Z_{pq}\Delta I_q \\ Z_{qp}\Delta I_p + Z_{qq}\Delta I_q \\ \vdots \\ Z_{np}\Delta I_p + Z_{nq}\Delta I_q \end{bmatrix} \tag{3.7}$$

(for assumed current injections at bus p and q).

Thus, for row p and q, we have

$$\Delta V_p = Z_{pp}\Delta I_p + Z_{pq}\Delta I_q \tag{3.8a}$$

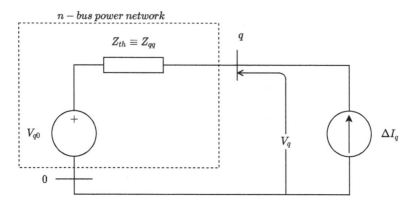

FIGURE 3.2   Thevenin's equivalent circuit at bus q.

$$\Delta V_q = Z_{qp}\Delta I_p + Z_{qq}\Delta I_q \tag{3.8b}$$

As $\Delta V_p = V_p - V_{p0}$ and $\Delta V_q = V_q - V_{q0}$, we can further write

$$V_p = V_{p0} + Z_{pp}\Delta I_p + Z_{pq}\Delta I_q \tag{3.9a}$$

$$V_q = V_{q0} + Z_{qp}\Delta I_p + Z_{qq}\Delta I_q \tag{3.9b}$$

Adding and subtracting $Z_{pq}\Delta I_p$ in Equation (3.9a) and $Z_{qp}\Delta I_q$ in Equation (3.9b), we get

$$V_p = V_{p0} + \left(Z_{pp} - Z_{pq}\right)\Delta I_p + Z_{pq}\left(\Delta I_p + \Delta I_q\right) \tag{3.10a}$$

$$V_q = V_{q0} + \left(Z_{qq} - Z_{qp}\right)\Delta I_q + Z_{qp}\left(\Delta I_p + \Delta I_q\right) \tag{3.10b}$$

As $[Z_{\text{BUS}}]$ is *symmetrical* in power system networks, $Z_{pq} = Z_{qp}$. Thus, Equations (3.10a) and (3.10b) can be rewritten as

$$V_p = V_{p0} + \left(Z_{pp} - Z_{pq}\right)\Delta I_p + Z_{pq}\left(\Delta I_p + \Delta I_q\right)$$

$$\text{and}\;\; V_q = V_{q0} + \left(Z_{qq} - Z_{pq}\right)\Delta I_q + Z_{pq}\left(\Delta I_p + \Delta I_q\right)$$

The equivalent circuit is shown in Figure 3.3.

To find equivalent impedance (Z_{th}) between bus p and q, the current and voltage sources are deactivated. The circuit of Figure 3.3 reduces to that shown in Figure 3.4.

Obviously in Figure 3.4, the Thevenin's impedance between bus p and q is given by

$$Z_{th(p-q)} = \left(Z_{pp} + Z_{qq} - 2Z_{pq}\right) \tag{3.11}$$

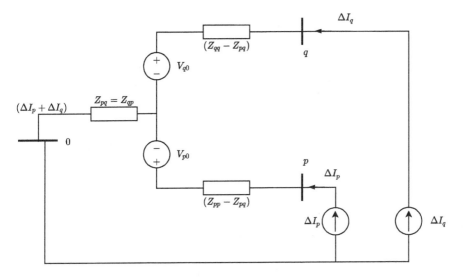

FIGURE 3.3 Thevenin's equivalent circuit of the p-th, q-th bus in the multi-bus power network.

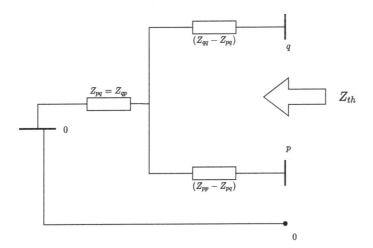

FIGURE 3.4 Determination of Thevenin's equivalent impedance between bus p and q.

Assuming a load impedance Z_l between bus p and q, the *Thevenin's current* through the load impedance Z_l is given by

$$I_l = \frac{V_{p0} - V_{q0}}{Z_{th(p-q)} + Z_l} \tag{3.12}$$

where I_l is the current through Z_l and $(V_{p0} - V_{q0})$ is the open circuit voltage between bus p and q.

3.3   ADDITION AND REMOVAL OF A LINE IN A POWER SYSTEM NETWORK

Let $V_1, V_2, \ldots, V_n$ = Bus voltages (in p.u) in the power system network

$I_1, I_2, \ldots, I_n$ = Known bus current injection (in p.u.) at respective buses.

Z_x and Z_y = p.u. impedances of lines to be added in the system between buses p-q and r-s, respectively.

I_x and I_y = Currents (in p.u.) in branches Z_x and Z_y, respectively.

$V_1', V_2', \ldots, V_n'$ = bus voltages (in p.u.) in the same power network after addition of Z_x and Z_y in the network.

Here,

$$[V] = \begin{bmatrix} V_1 \\ V_2 \\ \vdots \\ V_p \\ V_q \\ V_r \\ V_s \\ \vdots \\ V_n \end{bmatrix} ; [I] = \begin{bmatrix} I_1 \\ I_2 \\ \vdots \\ I_p \\ I_q \\ I_r \\ I_s \\ \vdots \\ I_n \end{bmatrix} ; [V'] = \begin{bmatrix} V_1' \\ V_2' \\ \vdots \\ V_p' \\ V_q' \\ V_r' \\ V_s' \\ \vdots \\ V_n' \end{bmatrix}$$

$$[Z_{\text{Bus}}] = \begin{bmatrix} Z_{11} & Z_{12} & \cdots & Z_{1p} & Z_{1q} & Z_{1r} & Z_{1s} & \cdots & Z_{1n} \\ Z_{21} & Z_{22} & \cdots & Z_{2p} & Z_{2q} & Z_{2r} & Z_{2s} & \cdots & Z_{2n} \\ \vdots & \vdots & & \vdots & \vdots & \vdots & \vdots & & \vdots \\ Z_{p1} & Z_{p2} & \cdots & Z_{pp} & Z_{pq} & Z_{pr} & Z_{ps} & \cdots & Z_{pn} \\ Z_{q1} & Z_{q2} & \cdots & Z_{qp} & Z_{qq} & Z_{qr} & Z_{qs} & \cdots & Z_{qn} \\ Z_{r1} & Z_{r1} & \cdots & Z_{rp} & Z_{rq} & Z_{rr} & Z_{rs} & \cdots & Z_{rn} \\ Z_{s1} & Z_{s2} & \cdots & Z_{sp} & Z_{sq} & Z_{sr} & Z_{ss} & \cdots & Z_{sn} \\ \vdots & \vdots & \cdots & \vdots & \vdots & \vdots & \vdots & \cdots & \vdots \\ Z_{n1} & Z_{n2} & \cdots & Z_{np} & Z_{nq} & Z_{nr} & Z_{ns} & \cdots & Z_{nn} \end{bmatrix}$$

Also, $[V] = [Z_{\text{BUS}}][I]$

Figure 3.5a represents the system in the initial state, while Figure 3.5b represents the system after Z_x and Z_y are added at the designated buses.

From Figure 3.5b, we have

$$Z_x I_x = V_p' - V_q' \quad \text{and} \quad Z_y I_y = V_r' - V_s' \tag{3.13}$$

Equation (3.13) can be represented in matrix form as shown in Equation (3.13a)

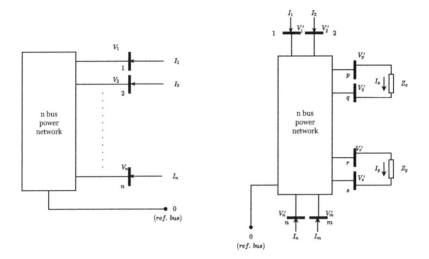

FIGURE 3.5 n-bus power system with addition of lines in designated buses.

$$\begin{bmatrix} Z_x & 0 \\ 0 & Z_y \end{bmatrix}\begin{bmatrix} I_x \\ I_y \end{bmatrix} = \begin{matrix} x \\ y \end{matrix}\begin{bmatrix} 0 & \cdots & 1 & -1 & 0 & 0 & \cdots & 0 \\ 0 & \cdots & 0 & 0 & 1 & -1 & \cdots & 0 \end{bmatrix}\begin{bmatrix} V_1' \\ V_2' \\ \vdots \\ V_p' \\ V_q' \\ V_r' \\ V_s' \\ \vdots \\ V_n' \end{bmatrix}$$

(3.13 a)

i.e., $\begin{bmatrix} Z_x & 0 \\ 0 & Z_y \end{bmatrix}\begin{bmatrix} I_x \\ I_y \end{bmatrix} = [A_C][V']$ (A_C being the connection matrix) (3.13 b)

As $(-I_x)$ enters bus p while $(+I_x)$ enters bus q and $(-I_y)$ enters bus r while $(+I_y)$ enters bus s in Figure 3.5b, we can express the current vector $[I_\Delta]$ as

$$[I_\Delta] = \begin{bmatrix} 0 \\ \vdots \\ -I_x \\ I_x \\ -I_y \\ I_y \\ \vdots \\ 0 \end{bmatrix}\begin{matrix} 1 \\ \vdots \\ p \\ q \\ r \\ s \\ \vdots \\ n \end{matrix}\begin{bmatrix} 0 & 0 \\ \vdots & \vdots \\ -1 & 0 \\ 1 & 0 \\ 0 & -1 \\ 0 & 1 \\ \vdots & \vdots \\ 0 & 0 \end{bmatrix}\begin{bmatrix} I_x \\ I_y \end{bmatrix} = -[A_C^T]\begin{bmatrix} I_x \\ I_y \end{bmatrix}$$

(3.14)

Change in bus voltages can be obtained using Equation (3.14)

$$[\Delta V] = [V'] - [V] = [Z_{\text{BUS}}][I_\Delta] = -[Z_{\text{BUS}}]\left[A_C^T\right]\begin{bmatrix} I_x \\ I_y \end{bmatrix} \tag{3.15}$$

Also, $[V'] = [V] + [\Delta V] = [V] + [Z_{\text{BUS}}][I_\Delta]$

$$[V'] = [V] - [Z_{\text{BUS}}]\left[A_C^T\right]\begin{bmatrix} I_x \\ I_y \end{bmatrix} \tag{3.16}$$

Substituting Equation (3.16) in (3.13) yields

$$\begin{bmatrix} Z_x & 0 \\ 0 & Z_y \end{bmatrix}\begin{bmatrix} I_x \\ I_y \end{bmatrix} = [A_C][V'] = [A_C]\left\{[V] - [Z_{\text{BUS}}]\left[A_C^T\right]\begin{bmatrix} I_x \\ I_y \end{bmatrix}\right\}$$

$$= [A_C][V] - [A_C][Z_{\text{BUS}}]\left[A_C^T\right]\begin{bmatrix} I_x \\ I_y \end{bmatrix}$$

or,
$$\left[\begin{bmatrix} Z_x & 0 \\ 0 & Z_y \end{bmatrix} + [A_C][Z_{\text{BUS}}]\left[A_C^T\right]\right]\begin{bmatrix} I_x \\ I_y \end{bmatrix} = [A_C][V] = \begin{bmatrix} V_p - V_q \\ V_r - V_s \end{bmatrix}$$

$$\tag{3.17}$$

or,
$$\begin{bmatrix} I_x \\ I_y \end{bmatrix} = \left[\begin{bmatrix} Z_x & 0 \\ 0 & Z_y \end{bmatrix} + [A_C][Z_{\text{BUS}}]\left[A_C^T\right]\right]^{-1}[A_C][V]$$

or,
$$\begin{bmatrix} I_x \\ I_y \end{bmatrix} = [Z]^{-1}[A_C][V] = [Z]^{-1}[A_C]\begin{bmatrix} V_p - V_q \\ V_r - V_s \end{bmatrix} \tag{3.18}$$

where
$$[Z] = \left[\begin{bmatrix} Z_x & 0 \\ 0 & Z_y \end{bmatrix} + [A_C][Z_{\text{BUS}}]\left[A_C^T\right]\right]$$

$(V_p - V_q))$ and $(V_r - V_s)$ are the *open circuit voltages* between buses (p-q) and (r-s), respectively In the original power network (Figure 3.5a) and as evident from Equation (3.18),

$$[A_C][Z_{BUS}][A_C^T] = \begin{array}{c} x \\ y \end{array} \begin{bmatrix} 1 & -1 & 0 & 0 \\ 0 & 0 & 1 & -1 \end{bmatrix}$$

$$\times \begin{bmatrix} Z_{pp} & Z_{pq} & Z_{pr} & Z_{ps} \\ Z_{pq} & Z_{qq} & Z_{qr} & Z_{qs} \\ Z_{pr} & Z_{rq} & Z_{rr} & Z_{rs} \\ Z_{ps} & Z_{sq} & Z_{sr} & Z_{ss} \end{bmatrix} \begin{bmatrix} 1 & 0 \\ -1 & 0 \\ 0 & 1 \\ 0 & -1 \end{bmatrix}$$

(3.19)

$$= \begin{bmatrix} (Z_{pp} - Z_{pq}) - (Z_{qp} - Z_{qq}) & (Z_{pr} - Z_{ps}) - (Z_{qr} - Z_{qs}) \\ (Z_{rp} - Z_{rq}) - (Z_{sp} - Z_{sq}) & (Z_{rr} - Z_{rs}) - (Z_{sr} - Z_{ss}) \end{bmatrix}$$

However, from Equation (3.11) the Thevenin's impedance can be represented as

$$Z_{th(p-q)} = (Z_{pp} + Z_{qq} - 2Z_{pq}) \tag{3.20a}$$

$$\text{and, } Z_{th(r-s)} = (Z_{rr} + Z_{ss} - 2Z_{rs}) \tag{3.20 b}$$

Finally, substituting Equation (3.19) in (3.17), we can have

$$\begin{bmatrix} \dfrac{(Z_{pp} - Z_{pq}) - (Z_{qp} - Z_{qq}) + Z_x}{(Z_{rp} - Z_{rq}) - (Z_{sp} - Z_{sq})} & \dfrac{(Z_{pr} - Z_{ps}) - (Z_{qr} - Z_{qs})}{(Z_{rr} - Z_{rs}) - (Z_{sr} - Z_{ss}) + Z_y} \end{bmatrix} \begin{bmatrix} I_x \\ I_y \end{bmatrix}$$

$$= \begin{bmatrix} V_p - V_q \\ V_r - V_s \end{bmatrix} \tag{3.21}$$

I_x and I_y can be obtained from Equation (3.21) to observe the addition of impedances Z_x and Z_y between respective buses in the system. The removal of the line impedances Z_x and Z_y from the original network in Figure (3.5a) can be accomplished in a similar manner by treating the removals as additions of negative impedances $(-Z_x)$ and $(-Z_y)$ in Figure 3.5b and Equation (3.21).

Example 3.1

In a four-bus power system, the bus voltages are given by
$V_1 = 1.00\angle 0°$ p.u.; $V_2 = 0.99\angle 0°$ p.u., $V_3 = 0.98\angle 0°$ p.u. and $V_4 = 0.97\angle 0°$ p.u.
The $[Z_{BUS}]$ matrix is given by

$$[Z_{BUS}] = \begin{bmatrix} j0.28 & j0.25 & j0.20 & j0.15 \\ j0.25 & j0.22 & j0.15 & j0.15 \\ j0.20 & j0.15 & j0.20 & j0.12 \\ j0.15 & j0.15 & j0.12 & j0.20 \end{bmatrix} \text{p.u.}$$

Two lines Z_x and Z_y of p.u. reactances $j0.08$ and $j1$ p.u. are connected between buses 2–3 an 3–4, respectively. Find currents I_x and I_y flowing through line impedances Z_x and Z_y. Also, find the changes in bus voltages.

Solution

Letting bus nomenclatures $p = 2$, $q = 3$, $r = 3$ and $s = 4$ from Equation (3.21), we have

$$
\begin{bmatrix}
(Z_{22}-Z_{23})-(Z_{32}-Z_{33})+Z_x & (Z_{23}-Z_{24})-(Z_{33}-Z_{34}) \\
(Z_{32}-Z_{33})-(Z_{42}-Z_{43}) & (Z_{33}-Z_{34})-(Z_{43}-Z_{44})+Z_y
\end{bmatrix}
\begin{bmatrix} I_x \\ I_y \end{bmatrix}
=
\begin{bmatrix} V_2-V_3 \\ V_3-V_4 \end{bmatrix}
$$

$$
\begin{bmatrix}
(j0.22-j0.15)-(j0.15-j0.20)+j0.08 & (j0.15-j0.15)-(j0.20-j0.12) \\
(j0.15-j0.2)-(j0.15-j0.12) & (j0.2-j0.12)-(j0.12-j0.2)+j1
\end{bmatrix}
$$

$$
\times
\begin{bmatrix} I_x \\ I_y \end{bmatrix}
=
\begin{bmatrix}
0.99\angle0^\circ - 0.98\angle0^\circ \\
0.98\angle0^\circ - 0.97\angle0^\circ
\end{bmatrix}
$$

$$
\text{or,}\quad
\begin{bmatrix}
(j0.07+j0.05+j0.08) & (0-j0.08) \\
(-j0.05-j0.03) & (j0.08+j0.08+j1)
\end{bmatrix}
\begin{bmatrix} I_x \\ I_y \end{bmatrix}
=
\begin{bmatrix} 0.01 \\ 0.01 \end{bmatrix}
$$

$$
\begin{bmatrix} I_x \\ I_y \end{bmatrix}
=
\begin{bmatrix} j0.2 & -j0.08 \\ -j0.08 & j1.16 \end{bmatrix}^{-1}
\begin{bmatrix} 0.01 \\ 0.01 \end{bmatrix}
=
\begin{bmatrix} -j5.14 & -j0.35 \\ -j0.35 & -j0.88 \end{bmatrix}
\begin{bmatrix} 0.01 \\ 0.01 \end{bmatrix}
$$

$$
=
\begin{bmatrix} -j0.055 \\ -j0.012 \end{bmatrix}\text{p.u.}
\tag{3.21a}
$$

The change in bus voltages can be calculated as

$$
[\Delta V] = -I_x\left[Z_{\text{BUS}}^{p-q}\right] - I_y\left[Z_{\text{BUS}}^{(r-s)}\right]
$$

$$
\text{where,}\quad
\left[Z_{\text{BUS}}^{(p-q)}\right]=\left[Z_{\text{BUS}}^{(2-3)}\right]=
\begin{bmatrix}
Z_{1p}-Z_{1q} \\
Z_{2p}-Z_{2q} \\
\vdots \\
Z_{pp}-Z_{pq} \\
Z_{qp}-Z_{qq} \\
\vdots \\
Z_{np}-Z_{nq}
\end{bmatrix}
=
\begin{bmatrix}
Z_{12}-Z_{13} \\
Z_{22}-Z_{23} \\
Z_{32}-Z_{33} \\
Z_{42}-Z_{43}
\end{bmatrix}
$$

$$\therefore \left[Z_{BUS}^{(2-3)} \right] = \begin{bmatrix} j0.25 - j0.20 \\ j0.22 - j0.15 \\ j0.15 - j0.20 \\ j0.15 - j0.12 \end{bmatrix} = \begin{bmatrix} j0.05 \\ j0.07 \\ -j0.05 \\ j0.03 \end{bmatrix}$$

$$\text{Also, } \left[Z_{BUS}^{(r-s)} \right] = \left[Z_{BUS}^{(3-4)} \right] = \begin{bmatrix} Z_{1r} - Z_{1s} \\ Z_{2r} - Z_{2s} \\ \vdots \\ Z_{pr} - Z_{ps} \\ Z_{qr} - Z_{qs} \\ Z_{rr} - Z_{rs} \\ Z_{sr} - Z_{ss} \\ \vdots \\ Z_{nr} - Z_{ns} \end{bmatrix} = \begin{bmatrix} Z_{13} - Z_{14} \\ Z_{23} - Z_{24} \\ Z_{33} - Z_{34} \\ Z_{43} - Z_{44} \end{bmatrix}$$

$$= \begin{bmatrix} j0.20 - j0.15 \\ j0.15 - j0.15 \\ j0.20 - j0.12 \\ j0.12 - j0.20 \end{bmatrix} = \begin{bmatrix} j0.05 \\ j0 \\ j0.08 \\ -j0.08 \end{bmatrix}$$

$$\therefore \begin{bmatrix} \Delta V_1 \\ \Delta V_2 \\ \Delta V_3 \\ \Delta V_4 \end{bmatrix} = -I_x \begin{bmatrix} j0.05 \\ j0.07 \\ -j0.05 \\ j0.03 \end{bmatrix} - I_y \begin{bmatrix} j0.05 \\ j0 \\ j0.08 \\ -j0.08 \end{bmatrix}$$

Substituting the values of I_x and I_y from Equation (3.21a), the bus voltage changes are obtained as

$$\begin{bmatrix} \Delta V_1 \\ \Delta V_2 \\ \Delta V_3 \\ \Delta V_4 \end{bmatrix} = \begin{bmatrix} -0.02756 \\ -0.0038 \\ +0.0266 \\ -0.0007 \end{bmatrix}$$

3.4 CURRENT DISTRIBUTION FACTOR (Ψ_i) AND LINE OUTAGE FACTOR (λ)

Let us assume the current injection at bus r is altered in a multi-bus power network. This will cause change in bus voltages in the system as

$$\text{i.e., } [\Delta V] = [V'] - [V] = [Z_{BUS}][I_\Delta]$$

$$\text{or,} \quad \begin{bmatrix} \Delta V_1 \\ \Delta V_2 \\ \vdots \\ \Delta V_p \\ \Delta V_q \\ \vdots \\ \Delta V_n \end{bmatrix} = \begin{bmatrix} V_1' - V_1 \\ V_2' - V_2 \\ \vdots \\ V_p' - V_p \\ V_q' - V_q \\ \vdots \\ V_n' - V_n \end{bmatrix} = [Z_{\text{BUS}}] \begin{bmatrix} 0 \\ 0 \\ \vdots \\ \vdots \\ \Delta I_r \\ \vdots \\ 0 \end{bmatrix} \quad \text{[column } r \text{ of } Z_{\text{BUS}}] \, \Delta I_r \quad (3.22)$$

Here, we assumed nonprimed voltages to be the bus voltages before any current injection and is available from load flow study. The primed voltage quantities are respective bus voltages following current injection (ΔI_r) at bus r only. The current injection is constant, and for each line the line changing capacitance and any shunt connection to the line are omitted. All bus voltages are with respect to the reference bus.

Moreover, we have

$$\Delta V_p = Z_{pr} \Delta I_r$$

$$\text{and,} \quad \Delta V_q = Z_{qr} \Delta I_r \quad (3.23)$$

Let us now assume that Z_{pq} is the p.u. impedance of line between buses p and q. The change of line current in Z_{pq} can then be computed as

$$\Delta I_{pq} = \frac{\Delta V_p - \Delta V_q}{Z_{pq}} = \frac{Z_{pr}\Delta I_r - Z_{qr}\Delta I_r}{Z_{pq}} = \Delta I_r \left[\frac{Z_{pr} - Z_{qr}}{Z_{pq}} \right]$$

$$\therefore \frac{\Delta I_{pq}}{\Delta I_r} = \frac{Z_{pr} - Z_{qr}}{Z_{pq}} \quad (3.24)$$

Conventionally, the ratio $\left(\dfrac{\Delta I_{pq}}{\Delta I_r} \right)$ is termed as *current injection distribution factor* (Ψ_i).

$$\text{i.e.,} \quad \frac{\Delta I_{pq}}{\Delta I_r} = \Psi_i$$

$$\text{or,} \quad \Delta I_{pq} = \Psi_{i(pq)r} \Delta I_r \quad (3.25)$$

In a power system network, often the load increases in a bus leading to line loading beyond permissible value, and then it is required to reschedule the line loadings to keep the line loading within safe values. Mathematically, this is equivalent to change of current injection by ΔI_m and ΔI_n at buses m and n for reducing the line loading of line between buses m-n. Equation (3.25) can be rewritten as

$$\Delta I_{pq} = \Psi_{i(pq)m} \Delta I_m + \Psi_{i(pq)n} \Delta I_n \quad (3.26)$$

The next concept we would like to introduce is the line outage distribution factor (λ).

$$\lambda_{pq(m-n)} = \frac{\Delta I_{pq}}{I_{mn}}$$

(3.27)

$$\text{and, } \lambda_{rs(m-n)} = \frac{\Delta I_{rs}}{I_{mn}}$$

Clearly, the post contingency line currents in lines p-q and r-s are

$$I'_{pq} = I_{pq} + \Delta I_{pq} = I_{pq} + \lambda_{pq(m-n)} I_{mn}$$

(3.28a)

$$\text{and } I'_{rs} = I_{rs} + \Delta I_{rs} = I_{rs} + \lambda_{rs(m-n)} I_{mn}$$

(3.28b)

Thus, it is now evident that if the pre-contingency line currents are known from usual load flow as well as the line outage distribution factors λ_{pq} and λ_{rs} for tripping a line m-n are known, we can obtain the values of post-contingency line currents I_{pq} and I'_{rs} in the healthy lines p-q and r-s.

3.5 SINGLE LINE CONTINGENCY

Let us assume that a line L_{mn} is tripped between buses m and n. Following the concepts developed in the preceding chapters, the removal of the line L_{mn} my be considered to be adding an impedance $(-Z_x)$ between buses m and n. Obviously it is assumed that the magnitude of the imaginary impedance $(-Z_x)$ is identical to the magnitude of the impedance of the line L_{mn} removed.

Figure 3.6 represents the Thevenin's equivalent circuit where (I_x) is the current flowing through $(-Z_x)$. Here,

$$I_x = \frac{V_m - V_n}{\left(Z_{mm} + Z_{nn} - 2\,Z_{mn}\right) + \left(-Z_x\right)}$$

(3.29)

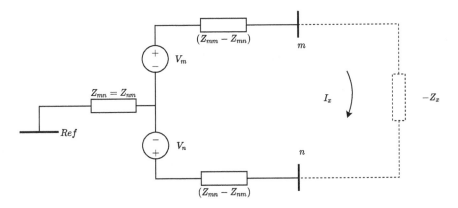

FIGURE 3.6 Removal of a line between buses m-n is equivalent to addition of $(-Z_x)$ impedance between buses m-n in Thevenin's equivalent circuit.

Here (V_m) and (V_n) are the pre-contingency voltages at buses m and n, respectively.

On the basis of the analytical reasoning in the preceding articles, we can say that the addition of impedance $(-Z_x)$ between buses m-n would cause current $\Delta I_m \ (=-I_x)$ in bus m and $\Delta I_n \ (=I_x)$ in bus n. Thus, the resulting current variation in the line between buses p-q can be expressed by the Equation (3.30).

$$\Delta I_{pq} = \Psi_{i(pq)m}\Delta I_m + \Psi_{i(pq)n}\Delta I_n$$

$$= \frac{Z_{pm}-Z_{qm}}{Z_{pq}}(-I_x) + \frac{Z_{pn}-Z_{qn}}{Z_{pq}}(I_x) \tag{3.30}$$

assuming Z_{pq} to be the impedance for line p-q

$$\text{or,} \quad \Delta I_{pq} = I_x \left[\frac{(Z_{pn}-Z_{pm})-(Z_{qn}-Z_{qm})}{Z_{pq}} \right] \tag{3.31a}$$

$$\Delta I_{pq} = \left(\frac{V_m - V_n}{Z_{mm}+Z_{nn}-2\ Z_{mn}-Z_x} \right) \left[\frac{(Z_{pn}-Z_{pm})-(Z_{qn}-Z_{qm})}{Z_{pq}} \right] \tag{3.31b}$$

However, during the state of pre-contingency, when the line L_{mn} was not removed, the current through this line (i.e. the current through the impedance $(-Z_x)$) was expressed as

$$I_x^{(0)} = \frac{V_m - V_n}{Z_x} \tag{3.32}$$

$I_x^{(0)}$ is the current through the line $L_{mn} \ (=Z_x)$ in the pre-contingency state, while I_x is the current through $(-Z_x)$ following line removal.

It may be noted here that $I_x^{(0)} \neq I_x$

Substituting Equation (3.32) in (3.31), we get

$$\Delta I_{pq} = \left[\frac{(Z_{pn}-Z_{pm})-(Z_{qn}-Z_{qm})}{Z_{pq}} \right] \frac{I_x^{(0)} Z_x}{Z_{mm}+Z_{nn}-2\ Z_{mn}-Z_x}$$

$$= \left(\frac{Z_x}{Z_{pq}} \right) \left[\frac{(Z_{pn}-Z_{pm})-(Z_{qn}-Z_{qm})}{Z_{mm}+Z_{nn}-2\ Z_{mn}-Z_x} \right] I_x^{(0)}$$

$$\text{i.e.,} \quad \frac{\Delta I_{pq}}{I_x^{(0)}} = -\left(\frac{Z_x}{Z_{pq}} \right) \left[\frac{(Z_{pm}-Z_{pn})-(Z_{qm}-Z_{qn})}{Z_{mm}+Z_{nn}-2\ Z_{mn}-Z_x} \right] \tag{3.33}$$

Here, $\left(\dfrac{\Delta I_{pq}}{I_x^{(0)}} \right)$ is the line utage distribution factor $\lambda_{pq(m-n)}$

Then we can say from Equation (3.33)

$$\lambda_{pq(m-n)} = -\left(\frac{Z_x}{Z_{pq}}\right)\left[\frac{(Z_{pm}-Z_{pn})-(Z_{qm}-Z_{qn})}{Z_{mm}+Z_{nn}-2\,Z_{mn}-Z_x}\right] \tag{3.34}$$

In a similar manner, for removal of line L_{mn}, the effect on line between buses r and s can be represented as

$$\lambda_{rs(m-n)} = -\left(\frac{Z_x}{Z_{rs}}\right)\left[\frac{(Z_{rm}-Z_{rn})-(Z_{sm}-Z_{sn})}{Z_{mm}+Z_{nn}-2\,Z_{mn}-Z_x}\right] \tag{3.35}$$

Thus, from Equations (3.34) and (3.35), we can find new line currents I'_{pq} and I'_{rs} following a line contingency in the power system between buses m-n.

$$I'_{pq}-I_{pq}=\Delta I_{pq}=-\left(\frac{Z_x}{Z_{pq}}\right)\left[\frac{(Z_{pm}-Z_{pn})-(Z_{qm}-Z_{qn})}{Z_{mm}+Z_{nn}-2\,Z_{mn}-Z_x}\right]I_x^{(0)}$$

$$\text{i.e.,}\quad I'_{pq}=I_{pq}-\left(\frac{Z_x}{Z_{pq}}\right)\left[\frac{(Z_{pm}-Z_{pn})-(Z_{qm}-Z_{qn})}{Z_{mm}+Z_{nn}-2\,Z_{mn}-Z_x}\right]I_x^{(0)} \tag{3.36a}$$

$$\text{i.e.,}\quad I'_{pq}=I_{pq}+\lambda_{pq(m-n)}I_x^{(0)} \tag{3.36b}$$

$$\text{i.e.,}\quad I'_{rs}=I_{rs}-\left(\frac{Z_x}{Z_{rs}}\right)\left[\frac{(Z_{rm}-Z_{rn})-(Z_{sm}-Z_{sn})}{Z_{mm}+Z_{nn}-2Z_{mn}-Z_x}\right]I_x^{(0)} \tag{3.37a}$$

$$\text{i.e.,}\quad I'_{rs}=I_{rs}+\lambda_{rs(m-n)}I_x^{(0)} \tag{3.37b}$$

Example 3.2

Assume five bus power system. The $[Z_{BUS}]$ (in p.u.) is given by

$$[Z_{BUS}]=\begin{bmatrix} j20.52 & j20 & j20.09 & j20.25 & j20.15 \\ j20 & j20.85 & j20.05 & j20 & j20.58 \\ j20.09 & j20.05 & j20 & j20.61 & j20.56 \\ j20.25 & j20 & j20.61 & j20.89 & j20.03 \\ j20.15 & j20.58 & j20.56 & j20.63 & j20.09 \end{bmatrix}$$

Pre-contingency bus voltages are given as

$$V_1=(1+j0)\text{p.u.};\quad V_2=(0.99-j0.08)\text{p.u.};\quad V_3=(0.98-j0.085)\text{p.u.};$$
$$V_4=(0.975-j0.06)\text{p.u.};\quad V_5=(0.98-j0.05)\text{p.u.}$$

Assume removal of the line between buses 2 and 4. The series impedance of line between buses 2 and 4, that is, L_{34} are $j0.05$ and $j0.1$ p.u., respectively. Compute the new current in the line between buses 3 and 4. In addition, obtain the value of the line outage distribution factor of L_{34} for removal of line L_{24}.

Solution

Let $p = 3$; $q = 4$; $m = 2$; $n = 4$

From Equation (3.34), we can write

$$\lambda_{34(2-4)} = -\left(\frac{Z_{24(L)}}{Z_{34(L)}}\right)\left[\frac{(Z_{32}-Z_{34})-(Z_{42}-Z_{44})}{Z_{22}+Z_{44}-2\,Z_{24}-Z_{24}}\right]$$

$$= -\frac{j0.05}{j0.1}\left[\frac{(j20.05-j20.61)-(j20-j20.89)}{j20.85+j20.89-2(j20)-j0.05}\right]$$

$$= -0.5\left[\frac{-j0.56+j0.89}{j1.69}\right] = -0.0976$$

∴ Line outage distribution factor of line L_{34} for outage of line L_{24} has been obtained as (−0.0976).

$$\text{Also, } I_{34}\left(\text{pre contingency}\right) = \frac{V_3-V_4}{Z_{34(L)}} = \frac{(0.98-j0.085)-(0.975-j0.06)}{j0.1}$$

$$= \frac{0.005-j0.025}{j0.1} = -(0.25+j0.05)\text{p.u.}$$

$$I_{24}\left(\text{pre contingency}\right) = \frac{V_2-V_4}{Z_{24(L)}} = \frac{(0.99-j0.08)-(0.975-j0.06)}{j0.05}$$

$$= \frac{0.015-j0.02}{j0.05} = -(0.4+j0.3)\text{p.u.}$$

Following Equation (3.36) we now have, for new line current in L_{34},

$$I'_{34} = I_{34}-\lambda_{34(2-4)}I_{24} = -(0.25+j0.05)-\left\{(-0.0976)\left[-(0.4+j0.3)\right]\right\}$$

$$= (-0.21-j0.079)\text{p.u.} = -(0.21+j0.079)\text{p.u.}$$

3.6 CONTINGENCY ANALYSIS OF INTERCONNECTORS

Sub-grids in a large power system are interconnected by interconnectors (tie-lines), and the contingency analysis of such lines is important for the power system planning and operation. $[Z_{\text{BUS}}]$ technique is an effective tool in contingency analysis of interconnectors.

Figure 3.7 represents the schematic of two interconnectors being connected between two sub-grids A and B. Let the interconnector connecting bus p of area A to

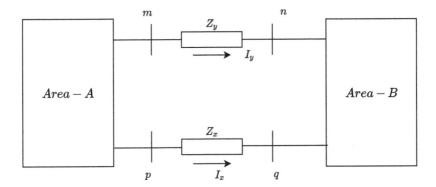

FIGURE 3.7 Two sub-grids A and B are interconnected through interconnections.

bus q of area B has impedance Z_x, while that connecting bus m of area A and bus n of area B has impedance Z_y.

First, let us assume that the sub-grids are not interconnected and $V_{1A}, V_{2A},\ldots, V_{mA}$ are the bus voltages for the buses 1, 2, ..., m in area A, while $V_{1B}, V_{2B},\ldots, V_{nB}$ are the bus voltages for the buses 1, 2, ..., n in area B. The respective bus impedance metrices of area A and area B are $[Z_{\mathrm{BUS}}]_A$ and $[Z_{\mathrm{BUS}}]_B$ during isolated operation of area A and area B. Moreover, the current injections in the respective buses are $I_{1A}, I_{2A},\ldots, I_{mA}$ for area A and $I_{1B}, I_{2B},\ldots, I_{nB}$ for area B.

Using the relation $[V] = [Z_{\mathrm{BUS}}]\,[I]$, we have for this system,

$$
\begin{bmatrix}
V_{1A} \\
V_{2A} \\
\vdots \\
V_{pA} \\
V_{mA} \\
\ldots \\
V_{1B} \\
V_{2B} \\
\vdots \\
V_{qB} \\
V_{nB}
\end{bmatrix}
=
\begin{bmatrix}
Z_A & 0 \\
0 & Z_B
\end{bmatrix}
\begin{bmatrix}
I_{1A} \\
I_{2A} \\
\vdots \\
I_{pA} \\
I_{mA} \\
\ldots \\
I_{1B} \\
I_{2B} \\
\vdots \\
I_{qB} \\
I_{nB}
\end{bmatrix}
\tag{3.38}
$$

where, $Z_A \equiv [Z_{\mathrm{BUS}}]_A$ and $Z_B \equiv [Z_{\mathrm{BUS}}]_B$

Let us now assume that the tie-lines are interconnected and the currents flowing through these lines are I_x and I_y, respectively. Let the current flows alter the respective bus voltages from V_{1A} to V'_{1A}, V_{2A} to V'_{2A} ... V_{pA} to V'_{pA}, and so on in area A and V_{1B} to V'_{1B}, V_{2B} to V'_{2B} ... V_{pB} to V'_{pB} and so on for area B. The connection matrix can then be expressed as

$$[A_C] = \begin{bmatrix} A_{CA} & A_{CB} \end{bmatrix}$$

$$= \begin{array}{c} x \\ y \end{array} \begin{bmatrix} 0 & 0 & \cdots & 1 & 0 & \vdots & 0 & 0 & \cdots & 1 & 0 \\ 0 & 0 & \cdots & 1 & 0 & \vdots & 0 & 0 & \cdots & 0 & -1 \end{bmatrix} \tag{3.39}$$

Following Equation (3.16), we have

$$\begin{bmatrix} V'_{1A} \\ V'_{2A} \\ \vdots \\ V'_{pA} \\ V'_{mA} \\ \cdots \\ V'_{1B} \\ V'_{2B} \\ \vdots \\ V'_{qB} \\ V'_{nB} \end{bmatrix} = \begin{bmatrix} V_{1A} \\ V_{2A} \\ \vdots \\ V_{pA} \\ V_{mA} \\ \cdots \\ V_{1B} \\ V_{2B} \\ \vdots \\ V_{qB} \\ V_{nB} \end{bmatrix} - \begin{bmatrix} Z_A & 0 \\ 0 & Z_B \end{bmatrix} \begin{bmatrix} A_C^T \end{bmatrix} \begin{bmatrix} I_x \\ I_y \end{bmatrix} \tag{3.40}$$

In Equation (3.21) letting bus r as m and bus s as n we can write

$$\begin{bmatrix} (Z_{pp} + Z_{qq} + Z_x) & \vdots & (Z_{pn} + Z_{qn}) \\ \cdots & \cdots & \cdots \\ (Z_{mp} + Z_{nq}) & \vdots & (Z_{mm} + Z_{nn} + Z_y) \end{bmatrix} \begin{bmatrix} I_x \\ I_y \end{bmatrix} = \begin{bmatrix} V_p - V_q \\ V_m - V_n \end{bmatrix} \tag{3.41}$$

Because there are no interconnections between buses p-n and m-q hence $Z_{pn} = Z_{np}$ and $Z_{mq} = Z_{qm}$ are nonexistent quantities.

$$\therefore \begin{bmatrix} I_x \\ I_y \end{bmatrix} = \begin{bmatrix} (Z_{pp} + Z_{qq} + Z_x) & \vdots & (Z_{pn} + Z_{qn}) \\ \cdots & \cdots & \cdots \\ (Z_{mp} + Z_{nq}) & \vdots & (Z_{mm} + Z_{nn} + Z_y) \end{bmatrix}^{-1} \begin{bmatrix} V_p - V_q \\ V_m - V_n \end{bmatrix} \tag{3.41a}$$

Here, $[Z] = \begin{bmatrix} (Z_{pp} + Z_{qq} + Z_x) & \vdots & (Z_{pn} + Z_{qn}) \\ \cdots & \cdots & \cdots \\ (Z_{mp} + Z_{nq}) & \vdots & (Z_{mm} + Z_{nn} + Z_y) \end{bmatrix}$

$$= \underbrace{\begin{bmatrix} Z_{pp} & Z_{pm} \\ Z_{mp} & Z_{mm} \end{bmatrix}}_{\text{Sub matrix for system-}A} + \begin{bmatrix} Z_x & 0 \\ 0 & Z_y \end{bmatrix} + \underbrace{\begin{bmatrix} Z_{qq} & Z_{qn} \\ Z_{nq} & Z_{nn} \end{bmatrix}}_{\text{Sub matrix for system-}B} \tag{3.42}$$

$[Z]$ is the $[Z_{BUS}]$ matrix for the system with interconnections connected. Moreover, the change in bus voltages can be obtained using Equation (3.15).

$$
\begin{bmatrix} \Delta V_{1A} \\ \vdots \\ \Delta V_{pA} \\ \Delta V_{mA} \end{bmatrix} = \begin{bmatrix} V'_{1A} \\ \vdots \\ V'_{pA} \\ V'_{mA} \end{bmatrix} - \begin{bmatrix} V_{1A} \\ \vdots \\ V_{pA} \\ V_{mA} \end{bmatrix} = -[Z_{BUS}]_A [A_{CA}^T] \begin{bmatrix} I_x \\ I_y \end{bmatrix} \tag{3.34a}
$$

and

$$
\begin{bmatrix} \Delta V_{1B} \\ \vdots \\ \Delta V_{pB} \\ \Delta V_{nB} \end{bmatrix} = \begin{bmatrix} V'_{1B} \\ \vdots \\ V'_{pB} \\ V'_{nB} \end{bmatrix} - \begin{bmatrix} V_{1B} \\ \vdots \\ V_{pB} \\ V_{nB} \end{bmatrix} = -[Z_{BUS}]_B [A_{CA}^T] \begin{bmatrix} I_x \\ I_y \end{bmatrix}
$$

Thus, the line currents can be obtained from Equation (3.41a), while the change in bus voltages are obtained from Equations (3.43a) and (3.43b).

3.7 CONTINGENCY ANALYSIS EMPLOYING DC POWER FLOW MODEL

In the *DC power flow* method, we assume that the power system is loss-less and each transmission line is represented by series reactance only; the sunt charging is neglected. Angular difference between voltages at adjacent buses $i-j$ are assumed to be small so that $\sin(\delta_i - \delta_j) \approx (\delta_i - \delta_j)$ rad; and $\cos \delta_i \approx \cos \delta_j$. It has been assumed that the system voltages at the buses are 1.00 p.u. The p.u. current flow from the i-th bus to the k-th bus through a series line reactance of (jx_l) p.u. is expressed as

$$
I_{i-j} = \frac{V_i - V_j}{jx_l} = \frac{|V_i|(\cos \delta_i + j \sin \delta_i) - |V_j|(\cos \delta_j + j \sin \delta_j)}{jx_l}
$$

$$
\text{or, } I_{i-j} = \frac{(\cos \delta_i - \cos \delta_j) + j(\sin \delta_i - \sin \delta_j)}{jx_l} = \frac{\delta_i - \delta_j}{x_l} \text{ p.u.} \tag{3.44}
$$

$$
\left(\because |V_i| = |V_j| = 1.00 \text{ p.u.} \right)
$$

$\therefore I_{i-j}$ becomes a real quantity. The p.u. power flow in the connecting line between bus $i-j$ is expressed as

$$
P_{i-j} = \frac{|V_i||V_j|}{x_l} \sin(\delta_i - \delta_j) \approx \frac{\delta_i - \delta_j}{x_l} \text{ p.u.} \tag{3.45}
$$

Comparing Equations (3.44) and (3.45) we see that once the voltages at buses of the power network are assumed to be 1.00 p.u., the line current and power flow lead to the same expression; hence, both are synonymous in the DC power flow method, that is, $I_{i-j} \equiv P_{i-j} = \dfrac{\delta_i - \delta_j}{x_l}$

ΔI_{i-j}, the change in line current is given by

$$\Delta I_{i-j} = \frac{\Delta(\delta_i - \delta_j)}{x_l} = \frac{\Delta\delta_i - \Delta\delta_j}{x_l} \text{ p.u} \tag{3.46}$$

Next, we look at Equation (3.27) to apply the concept of line outage factor $\lambda_{i-j(m-n)}$ for assessing the line outage factor λ for line $i - j$ due to assumed removal of line $m-n$ in the power system network. Following Equation (3.34) we can write

$$\lambda_{i-j(m-n)} = -\left(\frac{Z_x}{Z_{ij}}\right)\left[\frac{(Z_{im} - Z_{in}) - (Z_{jm} - Z_{jn})}{Z_{mm} + Z_{nn} - 2\,Z_{mn} - Z_x}\right]$$

$\therefore I'_{i-j}\left(\text{new flow of current in line } i - j \text{ following contingency at line } m - n\right)$

$$= I_{ij}\left(\text{i.e., pre contingency line current in } i - j\right) - \left(\frac{Z_x}{Z_{ij}}\right)\left[\frac{(Z_{im} - Z_{in}) - (Z_{jm} - Z_{jn})}{Z_{mm} + Z_{nn} - 2\,Z_{mn} - Z_x}\right]I_x^{(0)} \tag{3.47}$$

Where, for removal of line $m-n$, $Z_x \equiv Z_{mn}$ and

$$I_x^{(0)} = \frac{V_m - V_n}{Z_{mn}} \equiv I_x^{(0)}$$

Since $\lambda_{i-j(m-n)} = \dfrac{\Delta I_{ij}}{I_{mn}}$ (see equation 3.27), in dc pwer flow model

$$\lambda_{i-j(m-n)} = \frac{\Delta P_{ij}}{P_{mn}} = -\left(\frac{Z_x}{Z_{ij}}\right)\left[\frac{(Z_{im} - Z_{in}) - (Z_{jm} - Z_{jn})}{Z_{mm} + Z_{nn} - 2\,Z_{mn} - Z_x}\right] \tag{3.48}$$

3.8   CONCEPT OF EQUIVALENCING IN POWER SYSTEM NETWORK

In a complex power system, different areas in the power system network are interconnected through tie-lines. Once the contingency analysis is done for one system, the other system (i.e. the external system network) needs to be represented by a reduced equivalent network so that the modified bus impedance matrix is available.

Let us assume that tie-line $T-1$ interconnects bus i of area 1 to bus j of area 2, as shown in Figure 3.8. Let $[Y_{\text{BUS}}]_1$ and $[Y_{\text{BUS}}]_2$ be the respective bus admittance matrices. The overall $[Y_{\text{BUS}}]$ of the entire system is expressed as

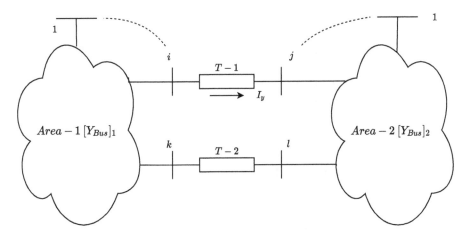

FIGURE 3.8 Two areas of power system interconnected by tie-lines.

$$[Y_{\text{BUS}}] = \begin{array}{c} k \\ i \\ \vdots \\ \vdots \\ l \\ j \\ \vdots \\ 1 \end{array} \left[\begin{array}{cccc:cccc} & & & & \vdots & Y_{kl} & 0 & \cdots & 0 \\ & & & & \vdots & 0 & Y_{ij} & \cdots & 0 \\ & & & & \vdots & \vdots & \vdots & \cdots & \vdots \\ & & & & \vdots & 0 & 0 & \cdots & 0 \\ \cdots & \cdots & \cdots & \cdots & \vdots & \cdots & \cdots & \cdots & \cdots \\ Y_{lk} & 0 & \cdots & 0 & \vdots & & & & \\ 0 & Y_{ji} & \cdots & 0 & \vdots & & [Y_{\text{BUS}}]_2 & & \\ \vdots & \vdots & \vdots & \cdots & \vdots & \vdots & & & \\ 0 & 0 & \cdots & 0 & \vdots & & & & \end{array} \right] \qquad (3.49)$$

Once the contingency analysis of area 2 is attempted, proper equivalencing of network at area 1 is required. We can use the popular Kron's network reduction technique and use the following expression

$$Y_{ij(\text{new})} = Y_{ij(\text{old})} - \frac{Y_{in} \times Y_{nj}}{Y_{nn}},$$

where n is the node number that is to be eliminated. Kron's network reduction technique is applied to $[Y_{\text{BUS}}]_1$. Let the new bus admittance matrix of area 1 be $[Y_{\text{BUS}}]'_1$ where all the nodes except nodes k and i are eliminated;

$$\text{i.e.,} \quad [Y_{\text{BUS}}]'_1 = \begin{bmatrix} Y'_{kk} & Y'_{ki} \\ Y'_{ik} & Y'_{ii} \end{bmatrix}$$

Equation (3.49) can now be modified as

$$[Y_{\text{BUS}}]_{\text{new}} = \begin{array}{c} k \\ i \\ \cdots \\ l \\ j \\ \vdots \\ 1 \end{array} \begin{bmatrix} Y'_{kk} & Y'_{ki} & \vdots & Y_{kl} & 0 & \cdots & 0 \\ Y'_{ik} & Y'_{ii} & \vdots & 0 & Y_{ij} & \cdots & 0 \\ \cdots & \cdots & \cdots & \cdots & \cdots & \cdots & \cdots \\ Y'_{lk} & 0 & \vdots & & & & \\ 0 & Y'_{ji} & \vdots & & [Y_{\text{BUS}}]_2 & & \\ \vdots & \vdots & \vdots & \vdots & & & \\ 0 & 0 & \vdots & & & & \end{bmatrix} \qquad (3.50)$$

The voltage current equation can then be written as

$$\begin{bmatrix} I'_k \\ I'_i \\ \cdots \\ I'_l \\ I'_j \\ \vdots \\ I'_1 \end{bmatrix} = \begin{bmatrix} Y'_{kk} & Y'_{ki} & \vdots & Y_{kl} & 0 & \cdots & 0 \\ Y'_{ik} & Y'_{ii} & \vdots & 0 & Y_{ij} & \cdots & 0 \\ \cdots & \cdots & \cdots & \cdots & \cdots & \cdots & \cdots \\ Y_{lk} & 0 & \vdots & & & & \\ 0 & Y_{ji} & \vdots & & [Y_{\text{BUS}}]_2 & & \\ \vdots & \vdots & \vdots & \vdots & & & \\ 0 & 0 & \vdots & & & & \end{bmatrix} \begin{bmatrix} V'_k \\ V'_i \\ \cdots \\ V'_l \\ V'_j \\ \vdots \\ V'_1 \end{bmatrix} \qquad (3.51)$$

In the next phase, bus k and i are eliminated. Kron's Node elimination technique is used to eliminate bus k and i of area 1 and to retain buses of area 2 only. This gives us

$$\begin{bmatrix} Y''_{ll} & Y''_{lj} & \cdots & Y''_{l1} \\ Y''_{jl} & Y''_{ljj} & \cdots & Y''_{j1} \\ \vdots & \vdots & & \vdots \\ Y''_{1l} & Y''_{1j} & \cdots & Y''_{11} \end{bmatrix} \begin{bmatrix} V'_l \\ V'_j \\ \vdots \\ V'_1 \end{bmatrix} = \begin{bmatrix} I'_l \\ I'_j \\ \vdots \\ I'_1 \end{bmatrix} \qquad (3.52)$$

The new $[Z_{\text{BUS}}]$ with respect to area 2 having equivalent representation of buses of area 1 is as shown below

$$[Z_{\text{BUS}}]'' = \begin{bmatrix} Y''_{ll} & Y''_{lj} & \cdots & Y''_{l1} \\ Y''_{jl} & Y''_{ljj} & \cdots & Y''_{j1} \\ \vdots & \vdots & & \vdots \\ Y''_{1l} & Y''_{1j} & \cdots & Y''_{11} \end{bmatrix}^{-1} \qquad (3.53)$$

while the new voltages for the buses in area 2 are given by

$$\begin{bmatrix} Z''_{ll} & Z''_{lj} & \cdots & Z''_{l1} \\ Z''_{jl} & Z''_{jj} & \cdots & Z''_{l1} \\ \vdots & \vdots & & \vdots \\ Z''_{1l} & Z''_{lj} & \cdots & Z''_{11} \end{bmatrix} \begin{bmatrix} I'_l \\ I'_j \\ \vdots \\ I'_1 \end{bmatrix} = \begin{bmatrix} V'_l \\ V'_j \\ \vdots \\ V'_1 \end{bmatrix} \qquad (3.54)$$

$$\text{i.e., } [V'] = [Z_{\text{BUS}}]'' [I']$$

The elements of $[Z_{\text{BUS}}]''$ has external network equivalents of area 1, and hence, it is possible to conduct the study of contingency analysis of the system in area 2 with equivalent network of area 1 embedded in the new $[Z_{\text{BUS}}]$. I_l' and I_j' are equivalent current injections at buses l and j, the boundary buses, while other bus current injections remain unchanged.

REFERENCE

1. A. Chakrabarti and S. Halder, *"Power System Analysis: Operation and Control"*, Third Edition, New Delhi: PHI Learning Pvt. Ltd.

4 Fundamental Concepts of Complex Network Theory

4.1 INTRODUCTION

Complex network theory has gained extensive approval and has been successfully applied in the assessment of complex systems such as computer networks, social interacting species, internet and many more. With the evolution of complex network theory, most complex systems in the universe can be abstracted as networks consisting of a set of edges connecting a set of vertices. The inherent structural features of these networks are assessed statistically using metrics. Earlier, the concept of complex network theory was applied in abstracted networks such as *random network* [1], *small-world networks* [2] and *scale-free networks* [3,4]. Scale-free networks are fragile to intentional attacks but robust against random failures of nodes [4]. They are more prone to cascading failure triggered by intentional attacks rather than random networks [5,6].

The complexity of power systems arises not just from the instant power balance of generators and consumers in large-scale transmission network across multitude of countries but also from the decision making of system operators to keep the system secure and reliable. Furthermore, there is a strong link between topological structure and operational performance in power systems. For instance, a large-scale blackout is more possible to be triggered by removing some critical buses or lines, which are essential elements of the topological structure of power systems. Consequently, power systems are naturally analysed under the framework of complex networks [7–9].

Despite the application of complex network theory in simulation of an electrical power system, it would not be appropriate until it is employed with power system-related constraints and features. Hence, it is more appropriate to utilise the concepts of complex network theory in assessing the structural vulnerability of electrical power grid associated with appropriate constraints and operational aspects related to electrical power system. Literature survey reveals that attempts have already been made to analyse the structural vulnerability in the North American [10] and European power grids [11–13]. In this chapter, the complex network theory has been employed to analyse its features related to vulnerability taking into account the associated parameters of power system such as electrical distance, line flow limit and power transmission distribution. These parameters are inserted through appropriate mathematical framework into the traditional complex network metrics such as degree, geodesic distance, efficiency and betweenness. The concept of extended topological metrics of electrical betweenness and netability have been simulated to assess the vulnerability of the power network. Furthermore, this chapter explains the theoretical and mathematical concepts of resilience originating from percolation threshold, incorporating the concept of preferential probability in modelling of cascading failures.

4.2 PURE TOPOLOGICAL APPROACH

In conventional complex network theory approach, there is a primary and essential set of centrality indices which are utilised for measuring the importance of a vertex or an edge in a network. These indices have been classified into three categories: the first being that the centrality of a vertex in a network determines how it is related to other vertices (*degree centrality*), the second one is based on the concept that central vertices stand between others (*betweenness centrality*) and the third, namely *efficiency*, quantifies the contribution of a network performance when subjected to removal of vertices or edges. To determine the criticality of the component and (or) networks, these metrics form the base of the pure topological method.

In the complex network theory, each bus of the power system, which may be a power source or a power sink, can be modelled as a *vertex* (or *node*), and each transmission line and transformer can be modelled as an *edge* (or *line*), in which power flow may be transmitted between its terminals. The actual ability of a power transmission system to perform properly depends on its topological structure, impedance and flow limits of its lines [14].

The vulnerability is the ability of a network continuing to provide key services during random failures or intentional attacks. Complex network theory provides a feasible way to study the vulnerability of power grids, which has drawn a link between the topological structure and the vulnerability of networks. There are some salient concepts that are required for the vulnerability analysis of power grid complex network.

4.2.1 GEODESIC DISTANCE

The number of lines in a path connecting nodes i and j is called the length of the path. A *geodesic path* [15] (or the shortest path) between i and j is the path connecting these nodes with the minimum length. The length of the geodesic path is the geodesic distance d_{ij} between i and j. If one is dealing with a weighted graph, the length of a path is the sum of the weights of the lines constituting that path.

4.2.2 AVERAGE SHORTEST PATH LENGTH

The distance d_{ij} between the node i and j is defined as the number of the edges which form the shortest path connecting node i and j. The average shortest path length L is defined [2] as the mean of geodesic lengths over all node couples:

$$L = \frac{1}{N(N-1)} \sum_{i,j=1}^{N} d_{ij} \tag{4.1}$$

In a purely topological model, a power network is considered to be a network composed of *vertices* (*buses*) connected by *edges* (*transmission lines*). In most cases, an unweighted and undirected network model is utilised. All the vertices and edges are considered to be identical, without differences in their quantitative features or directions. In an unweighted and undirected graph, the length of a path is the number of edges in a path connecting vertices i and j.

4.2.3 Degree (Connectivity)

The *degree* or *connectivity* [15] of a node is traditionally measured by its degree in an unweighted topological model or its strength in a weighted model. In an unweighted and undirected network model (according to traditional graph theory), the degree of a vertex i is the number of edges connected to it (or the number of vertices adjacent to it);

$$c_i = \sum_j l_{ij} \qquad (4.2)$$

where l_{ij} represents the number of lines connecting i and j.

In a weighted network model, connectivity can also be expressed by the strength measured as the sum of the weights of the corresponding edges:

$$s_i = \sum_j w_{ij} \qquad (4.3)$$

where w_{ij} represents the weight of the line connecting i and j.

Cumulative degree distribution being an essential feature of the topological configuration of a network, it can be suitably employed in assessing the vulnerability of a network when the cumulative degree distribution of a network follows a Poisson's distribution. It becomes a homogenous network where each node has the same degree. However, if the distribution is a power law or exponential, the corresponding network is heterogeneous, in which there are some vertices with higher degree than others.

The degree indicates the connectivity of a node in a network; if a node has higher connectivity, the node has more connections between other nodes and is more relevant. Therefore, degree could be treated as a metric to measure the criticality of the nodes in networks.

4.2.4 Distance and Efficiency

The power grid can be abstracted into the complex network with a graph composed of numbers of buses as a set of nodes and numbers of lines as a set of links [15]. The total number of nodes and links of the graph are N and L, respectively. The association of nodes with each other can be shown using an adjacency matrix. The walk of minimal length between two nodes is known as the shortest path or geodesic path, where a walk from node i to node j is an alternating sequence of nodes and edges (a sequence of adjacent nodes) that begins with i and ends with j. The length of the walk is defined as the number of edges in the sequence.

Efficiency of power grid can be expressed as the following equation [15]:

$$E = \frac{1}{N(N-1)} \sum_{i \neq j} \frac{1}{d_{ij}} \qquad (4.4)$$

where N is the number of nodes and d_{ij} is the geodesic distance between nodes i and j.

The concept of distance d_{ij} may be explained as the difficulty to transfer the relevant quantity between the nodes (i, j) of a network. *Distance* generally depends on the path that one follows and should be defined as a function of the characteristics of the lines along the path. The economic and technical difficulties for transfer of electrical power through a path depend on both the power flow through the lines and on their impedance. With the same impedance, higher power flow increases costs; with the same power flow, higher impedance increases costs. Consequently, the distance from node i to node j along path k is related not only to the impedance of each line of the path but also to the power flows through the lines of the path.

4.2.5 ELECTRICAL DISTANCE

The *electrical distance* [16] considers characteristics of the lines along the path joining buses. So, it can be defined as the equivalent impedance Z_{ij}, that is, the Thevenin impedance between the two nodes i and j [17].

$$Z_{ij} = z_{ii} + z_{jj} - 2z_{ij} \qquad (4.5)$$

where Z is the element of the bus impedance matrix. The electrical distance is used instead of geodesic distance when considering an extended topological approach.

$$d_{ij} = Z_{ij} \qquad (4.6)$$

4.2.6 ELECTRICAL CENTRALITY AND ITS MEASUREMENT

The calculation of *electrical centrality* is entirely based on Z_{bus}. Electrical centrality [18] is computed using Z_{bus} impedance matrix, which, in turn, can be either obtained from Y_{bus} matrix of the distribution system or by building up formulation of Z_{bus} algorithm. The latter is harder to implement but much more practical and faster for larger systems.

However, Kirchhoff's laws are followed for the calculation of this measure. As the physical topology of a distribution network may be either radial or ring type, the Y_{bus} matrix is sparse in nature. The impedance matrix Z_{bus} gives the electrical topology of the system by representing the strength of connection of every node to every other node in the network electrically. In this chapter, the approach is to directly build formulation of Z_{bus} algorithm without using Y_{bus}.

Considering only topological properties, geodesic distance is used to find path impedance in complex networks. The weights of lines may be considered as in a weighted graph or may not be considered as in an unweighted graph. It is obvious that the smaller the geodesic distance between a pair of nodes is, the more efficiently the energy transmits. Under normal operations, geodesic distance is replaced by electrical distance. Electrical distance considers characteristics of the lines along the path joining buses.

Let us call Z_{ij} the equivalent impedance of the circuit whose ends are node i and node j; V_{ij} is the voltage between i and j and I_i is the current injected at node i and extracted at node j ($I_i = -I_j$). As shown in Figure 4.1, the equivalent impedance is defined as

$$Z_{ij} = \frac{V_{ij}}{I_i} \tag{4.7}$$

Furthermore, let $I_i = 1$, $I_j = -1$ and $I_h = 0 \; \forall \; h \neq i, j$ (meaning that a unit current is injected at node i and extracted at j, while no current is extracted nor injected in other nodes), then the computation of equivalent impedance is as shown in Figure 4.1 and amounts to

$$Z_{ij} = \frac{V_{ij}}{I_i} = V_{ij} \Rightarrow Z_{ij} = V_i - V_j$$

$$= \left(Z_{ii} - Z_{ij} \right) - \left(Z_{ij} - Z_{jj} \right)$$

$$= Z_{ii} + Z_{jj} - 2Z_{ij}$$

where Z_{ij} is the i-th, j-th element of the impedance matrix.

The magnitude of the corresponding entry in the Z_{bus} matrix represents the electrical distance or geodesic distance between the nodes i and j. Hence, the smaller the value of Z_{ij}, the shorter is the electrical distance or geodesic distance. From electrical topology, some nodes are heavily connected to the rest of the network while others are sparsely connected. The size of the node determines their relative importance in the network with respect to their connectivity with the rest of the network. Larger the nodes, higher are their connectivity with the remaining nodes. Electrical topology consists of $\frac{N(N-1)}{2}$ connections, where N is the number of buses or nodes; however, $N - 1$ connections are actually shown to correspond with the physical topology, which consists of exactly $N - 1$ links.

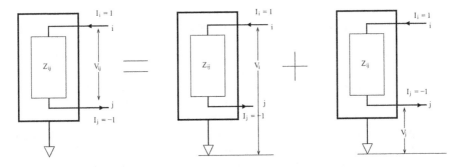

FIGURE 4.1 Equivalent impedance method.

4.2.7 BETWEENNESS

From the concept of complex network theory, it is evident that, if a vertex or edge participates in more number of paths, the corresponding component becomes more important for transmission in the entire network. Once it is assumed that the interactions or transmission is always through the shortest paths between two vertices, it can be used as an indication to justify the importance of a vertex or an edge in terms of its betweenness.

Energy transfer between two nonadjacent buses depends on the buses and lines of the geodesic paths connecting those buses. *Thus, the vulnerability measure of a network element can be determined by counting the number of geodesic paths going through it, and is defined as betweenness centrality of that element.*

Node betweenness of a node v can be expressed by the following formula [19]:

$$B(v) = \sum_i^N \sum_j^N \frac{\sigma_{ij}(v)}{\sigma_{ij}} \quad i \neq j \neq v \in V \tag{4.8}$$

where $\sigma_{ij}(v)$ is the number of geodesics from node i and node j through node v and σ_{ij} is the total number of geodesics between i and j. Similarly, the importance of lines in the power grid can be determined by edge betweenness. Mathematically, *line betweenness* centrality of a line can be obtained by the following formula [19]:

$$B(l) = \sum_i^N \sum_j^N \frac{\sigma_{ij}(l)}{\sigma_{ij}} \quad i \neq j \in V, l \in E \tag{4.9}$$

where $\sigma_{ij}(l)$ is the number of geodesics from node i to node j through line l. The betweenness centrality is based on the topological structure of a complex network.

A component in any network has a specific value of betweenness depending on the configuration of the network. When these values are higher, it indicates that a greater number of shortest paths pass through the component, highlighting the higher criticality of the component. Thus, critical components of a network can be identified by ranking the betweenness value of the network components.

4.3 EXTENDED TOPOLOGICAL APPROACH

The research on application of complex network theory during its early stage demonstrated some important concepts and measures that are suitable for various network types. Nonetheless, the features and physical aspects of different networks are naturally different from each other. Moreover, some particular features cannot be included with the general methodologies. If the complex network theory is mechanically applied to some fields without incorporating those particular features of these networks, results obtained would diverge from reality. For application of the methodologies of complex network theory in power systems, it is important to consider the electrical properties.

To apply centrality indices in power system calculations, it is important to redefine the corresponding metrics, particularly efficiency. It is convenient to express efficiency for the electrical network in terms of netability. Efficiency is redefined as netability in which line flow limit on each line and electrical distance are introduced into the efficiency index. Similarly, betweenness centrality is redefined as electrical betweenness by incorporating line flow limit on each line. *Power transfer distribution factor* (PTDF) is a matrix which is crucial for netability analysis as it reflects the line power flow sensitivity towards the change in the injected power of buses and withdrawn power at a reference bus.

In this chapter, four basic features of a power network have been considered that need to be extended using the topological approach. These features are incorporated to define the topological metrics into extended topological metrics.

- Bus Classification

    To avoid difficulties involved in differentiation and dynamic behaviour features of network components, all components have to be treated identically. Here, power flow is the characterising parameter which is considered to be transmitted from any vertex to any other. Generally, the buses in power transmission networks are considered as generation, transmission and load buses. Transmission is normally considered from generation to load buses.

- Line Flow Limits

    In a pure topological approach, edges are generally described in an unweighted manner to define related metrics such as distance, degree and betweenness. However, in power systems, line flow limits restrict the ability of power flow in a line due to congestion and economic load dispatch. These factors are important to assess the network security of the power system. These constraints need to be ascertained and applied for each line specifically as different lines may have distinct flow limits. To assess the vulnerability of the power network, electrical performance parameters are important to assess the network security.

- Flow-Based Network

    Transmission of power between two buses (vertices) is always supposed to be through the shortest path. This is the most unrealistic assumption from the viewpoint of electrical engineering. Power transmission from a generator bus to a load bus involves numerous lines which can be designated as paths having different extent of contribution. In simulation of power networks for a linear model of power flow, the various contributions of lines in the power network can be described by the PTDF.

    The conventional topological model in network theory describes the graph as unweighted and undirected. The identification of possible paths connecting two vertices is based on graph theory where transmission lines are assumed to be bidirectional. Some paths in undirected model may be not valid in the directed power transmission networks. PTDF characterises the behaviour of the system when the simulation between two vertices completely depends on physical rules. Because each element in PTDF is associated with a sign, the lines connecting to one vertex should be classified as input or output lines.

- Transmission capacity

    To maintain stable and secure operation, the capability of transferring power for each transmission line can be designated by its own transmission limit $P_{\max}^l$. In fact, for power transmission, not all lines reach their line flow limit at the same time. Once the line attains its power transmission limit, the power transmitted between corresponding pair of buses reaches its upper limit.

4.3.1   ELECTRICAL BETWEENNESS

In conventional topological approach, the concept of betweenness is defined as the sum of the probability for a vertex or an edge to belong to a randomly selected geodesic path linking any other pair of vertices. Betweenness is a more useful tool for measuring the magnitude of critical load on the given node/edge in the network. The importance of betweenness also includes network connectivity. It can be considered as a local metric to describe the degree centrality and corresponding criticality of components (vertices and edges) in complex networks.

Including the parameters of power grid, the extended betweenness of a bus v can be defined as [19]

$$T(v) = \frac{1}{2} \sum_{g \in G}^{N_G} \sum_{d \in D}^{N_D} C_g^d \sum_{l \in L^v} | f_l^{gd} | \tag{4.10}$$

where $v \neq g \neq d \in V$

$\sum_{l \in L^v} | f_l^{gd} |$ is the sum of PTDF of all the lines connecting bus v when power is injected at bus g and withdrawn at bus d.

$\frac{1}{2} C_g^d \sum_{l \in L^v} | f_l^{gd} |$ is the transmission power taken by bus v when power is transmitted from generator bus g to load bus d; G is the set of generation buses, D is the set of load buses, N_G is the number of generation buses, N_D is the number of load buses and L^v is the set of lines connecting bus v.

Similarly, the extended betweenness of a line l can be defined as follows [66]:

$$T(l) = \max \left[T^p(l), |T^n(l)| \right] l \in L \tag{4.11}$$

where T^p and T^n are the positive extended line betweenness and negative extended line betweenness of the line l.

$$T^p(l) = \sum_{g \in G} \sum_{d \in D} C_g^d f_l^{gd} \, g \neq d, f_l^{gd} > 0 \tag{4.12}$$

$$T^n(l) = \sum_{g \in G} \sum_{d \in D} C_g^d f_l^{gd} \, g \neq d, f_l^{gd} < 0 \tag{4.13}$$

where $C_g^d f_l^{gd}$ represents the power flowing on the line l when power is transmitted from g to d.

The concept of pure betweenness has been extended by introducing some electrical properties. The set of extended betweenness qualifies the contribution of a component to power transmission in the entire power grid, and in this respect, the elements of the power grid can be ranked according to their criticality.

4.3.2 VULNERABILITY USING GLOBAL EFFICIENCY

A general measure for vulnerability assessment of power grid is the so-called efficiency index, which can be used to describe the transmitting efficiency of the power grid. The transmitting efficiency of the link connecting nodes i and j is denoted by E, and the initial transmitting efficiency is 1 for every link in the network. The efficiency of a path between nodes i and j is the harmonic mean of efficiencies of all the links the path passed by. In all the paths connecting nodes i and j, the one with the maximal efficiency is called the most efficient path, and its efficiency is denoted as E. If there is no path between two nodes i and j, we have $d_{ij} = 0$.

As discussed in Equation (4.4), if there are N number of nodes and d_{ij} is the geodesic distance between nodes i and j, then the global efficiency of the network is expressed as [15]

$$E = \frac{1}{N(N-1)} \sum_{i \neq j} \frac{1}{d_{ij}} \tag{4.14}$$

Here, the sum is taken over all pairs of nodes of the unweighted and undirected network.

Vulnerability of a line in the network is a measure of drop in network functioning due to removal of that line in the system [14]. So, the damage of the line can be measured by the relative decrement in network performance. If E is the global efficiency of the network without any damage to the lines and $E(l)$ is the global efficiency of the network when line l is removed, then vulnerability of that line l can be calculated as [14]

$$V_e(l) = \frac{E - E(l)}{E} \tag{4.15}$$

The maximum of the vulnerabilities is vulnerability of the network;

$$V = \max\left[V_e(l)\right] m \tag{4.16}$$

Though the above method can be applied to power grid, some problems arise due to distinguished characteristics of power grid network as follows:

a) Geodesic distance considers only the shortest path between the nodes, but power can flow through any path between a pair of nodes.

b) There is no need to consider all pairs of nodes as power flows only from generation nodes to load nodes.

c) The transfer capabilities in transmitting power of a network are different which has not been taken into consideration in global efficiency calculation.

4.3.3 VULNERABILITY USING NETABILITY

Including general operating conditions such as power flow limit, transmission capacity and PTDF, the performance of a power grid can be evaluated using netability.

Power transmission capability C_{ij} is defined as the power injection at bus i when the first line of all the paths connecting the generation node i and the load node j reaches its limit. It is expressed as [19]

$$C_{ij} = \min_{l \in L} \left(\frac{P_{\max}^l}{|f_{ij}^l|} \right) \tag{4.17}$$

where $P_{\max}^l$ is the transmission limit of transmission line l and f_{ij}^l is the PTDF of line l of the path joining generation node i to load node j. This is the change of power on line l for injection at generation bus i and withdrawal at load bus j. f_{ij}^l is obtained as the difference between the entries f_{li} and f_{lj} of the PTDF matrix is calculated as [16]

$$A = H' B'^{-1} \tag{4.18}$$

where $B = N \times N$ admission matrix

$$B_{ij} = \frac{1}{x_{ij}}, i \neq j \tag{4.19}$$

$$B_{ii} = \sum_{j \neq i} \frac{1}{x_{ij}} \tag{4.20}$$

To avoid singularity, slack bus column and row are eliminated.

B = Sub-matrix of B where slack bus column and row are eliminated from B.

H = Transmission matrix of order $L \times N$

$$H_{li} = -H_{lj} = \frac{1}{x_{ij}} \tag{4.21}$$

$$H_{lk} = 0 \quad \forall k \neq i, j \tag{4.22}$$

L = Number of lines in the network joining the nodes.

N = Number of nodes in the network

H = Sub-matrix of H where slack bus column is eliminated from H.

Considering normal operation criteria, geodesic distance is replaced by electrical distance in netability model. Electrical distance considers characteristics of the lines along the path joining buses. So electrical distance is defined as the equivalent impedance Z_{ij} is the Thevenin impedance between the two nodes i and j as defined earlier $Z_{ij} = Z_{ii} + Z_{jj} - 2Z_{ij}$ where Z_{ij} is the i, jth element of the bus impedance matrix.

Hence, netability of a power grid network is expressed as

$$A = \frac{1}{N_G N_D} \sum_{i \in G} \sum_{j \in D} \frac{C_{ij}}{Z_{ij}} \tag{4.23}$$

where G = Set of generation buses
D = Set of load buses
N_G = Number of generation buses
N_D = Number of load buses
Line resistances are ignored.

Similarly, as in previous model, the vulnerability of line l is defined as the drop in netability of the network due to removal of the line l, that is,

$$V_a(l) = \frac{A - A(l)}{A} \tag{4.24}$$

where $A(l)$ is the netability of the network when line l is removed.

4.3.4 CRITICALITY ASSESSMENT USING [Z_{BUS}] CENTRALITY

A metric of electrical centrality that can be useful for installation of DG in sub-transmission system from the complex network theory viewpoint. It was proposed to differentiate the electrical topology and physical topology of power grid. This metric not only enables complex network analysis of power systems but is also more appropriate for the power grid than general topological analysis. This measure is based on Z_{bus} impedance matrix of a power system which finds more electrically central nodes in the system which is supposed to be highly connected to most other nodes for the placement of distributed generators in the sub-transmission system. Correct placement of DG in the transmission system is of strategic importance to improve the robustness of the grid. Incorrect placement may make the grid more vulnerable to failures or attacks.

The measure of *electrical centrality* (*electrical node significance*) is utilised to locate nodes where DGs can be placed. The size of the DGs depends on the relative importance of the nodes. The first step of the procedure is to calculate electrical centrality and significance of every node in the network. This calculation gives a clear indication of which nodes are more important in the network and offers a candidate set of locations where DGs can be placed. For all the test networks, most nodes indicated by electrical centrality differed from those indicated by electrical node significance. This shows that different methods can indicate multiple possibilities of locations for DG placement. Individually, they can produce slightly satisfactory results, but with a combination of these methods, the quality of results can be enhanced.

Electrical centrality is a measure used in topological analysis of power grid networks, which differentiates electrical structure of the grid from its topological structure. Electrical centrality uses the impedance matrix, or the Z bus matrix of the transmission system, to determine which nodes are more electrically central to the system and indicates them as candidate locations for the placement of DGs. The electrical topology of the grid indicates that power grids possess "electrical hubs," indicating that some nodes in the power grid have strong electrical connections with other parts of the network. This phenomenon is very different from the physical topology of the power grids as their average degree is usually between 2 and 5, indicating the absence of hubs. As mentioned in Section 4.2.6, electrical centrality is calculated using the Z bus matrix which is computed as the inverse of the Y bus matrix or the admittance matrix of the system. The Y bus matrix is usually sparse, and, hence, the Z bus matrix is obtained as a dense matrix. Every element in the matrix represents an equivalent electrical distance between two nodes. Since the Z bus matrix is a non-sparse, dense matrix, there are $\frac{N(N-1)}{2}$ different electrical connections possible.

REFERENCES

1. P. Erdos and A. Renyi, "On the evolution of random graphs", *Publications of the Mathematical Institute of the Hungarian Academy of Sciences*, vol. 5, pp. 17–61, 1960.
2. D.J. Watts and S.H. Strogatz, "Collective dynamics of 'small world' networks", *Nature*, vol. 393, pp. 440–442, 1998.
3. A.L. Barabasi and R. Albert, "Emergence of scaling in random networks", *Science*, vol. 286, pp. 509–512, 1999.
4. R. Albert, H. Jeong, and A.L. Barabási, "Error and attack tolerance of complex networks", *Nature*, vol. 406, pp. 378–382, 2000.
5. A.E. Motter and Y.C. Lai, "Cascade-based attacks on complex networks", *Physical Review E,* vol. 66, p. 065102, 2002.
6. P. Crucitti, V. Latora, and M. Marchiori, "Model for cascading failures in complexnetworks", *Physical Review E*, vol. 69, p. 045104, 2004.
7. R. Albert, I. Albert, and G.L. Nakarado, "Structural vulnerability of the North American power grid", *Physical Review E*, vol. 69, p. 025103, 2004.
8. V. Rosato, S. Bologna, and F. Tiriticco, "Topological properties of high-voltage electrical transmission networks", *Electric Power Systems Research*, vol. 77, pp. 99–105, 2007.
9. M. Rosas-Casals, S. Valverde, and R. Sol_e, "Topological vulnerability of the European power grid under errors and attacks", *International Journal of Bifurcation and Chaos*, vol. 17, no. 7, pp. 2465–2475, 2007.
10. R. Kinney, P. Crucitti, R. Albert, and V. Latora, "Modelling cascading failures in North American power grid", *European Physical Journal B*, vol. 46, pp. 101–107, 2005.
11. P. Crucitti, V. Latora, and M. Marchiori. "A topological analysis of the Italian electric power grid", *Physica A*, vol. 338, pp. 92–97, 2004.
12. P. Crucitti, V. Latora, and M. Marchiori, "Locating critical lines in high-voltage electrical power grids", *Fluctuation and Noise Letters*, vol. 5, pp. L201–L208, 2005.
13. M.E.J. Newman, "The structure and function of complex network", *SIAM Review*, vol. 45, pp. 167–256, 2003.
14. V. Latora and M. Marchiori, "Vulnerability and protection of infrastructure networks", *Physical Review E*, vol. 71, p. 15103, 2005.

15. V. Latora and M. Marchiori, "Efficient behaviour of small-world networks", *Physical Review Letter*, vol. 87, p. 198701, 2001.
16. S. Arianos, E. Bompard, A. Carbone, and F. Xue, "Power grid vulnerability: A complex network approach", *Chaos*, vol. 19, pp. 1–6, 2009.
17. J.J. Grainger and W.D. Stevenson, Jr., *Power System Analysis*, India: Tata McGraw-Hill, pp. 287–292, 1994.
18. S. Pahwa, D. Weerasinghe, C. Scoglio, and R. Miller, "A complex networks approach for sizing and siting of distributed generators in the distribution system", 45th North American Power Symposium, Manhattan KA, September 2013.
19. E. Bompard, D. Wu, and F. Xue, Structural vulnerability of power systems: A topological approach. *Electric Power Systems Research*, vol. 81, no. 7, pp. 1334–1340, 2011.

5 Vulnerability Assessment of Power Transmission Network

5.1 INTRODUCTION

Power generation, transmission and distribution, as an integrated system, is one of the critical infrastructures as it is widely distributed in a vast geographical region and indispensable to the modern world. Accidental failures and intentional attacks can lead to disastrous social and economic consequences. Therefore, to ensure reliable performance, electrical operation engineers need to assess the vulnerability of power grid networks and identify the critical elements that need essential back-up protection to ensure a more robust electrical system against natural or malicious threats.

The vastness, size of components and the complex dynamics of various components of power grids make them typical complex networks. Numerous researches including basic characteristics and steady state and transient state performance have been analysed since long, and in the present scenario of planned and unplanned attacks on civilisation and society have raised the issue of vulnerability of a power grid. Interestingly, the close relationship between topological structure and operation performance in power systems as a change in topological structure can alter the operational condition of a power system leading to change in its operational performance. Hence, there is an increasing interest in topological vulnerability assessment of power grid networks using complex network methodology.

In this chapter attempts have been made to explore the applicability of complex network theory in power distribution and transmission systems in assessment of vulnerability of any electrical power network.

In general, a vulnerable system operates with a "reduced level of security that renders it vulnerable to the cumulative effects of a series of moderate disturbances" [1]. From the viewpoint of power system operation, vulnerability is an estimate of the fragility of the system following a sequence of occurrences which might involve line or generator failure, disruptions, interruptions, blackout, breakdown or undesirable operations of protective relays, information or communication system malfunctioning and human errors.

Vulnerability assessment includes identification and quantification following the prioritising (or ranking) sequence while the system is under the threat of planned or unplanned contingencies or attacks [2]. In conventional methods of security assessment and contingency analysis, power system engineers use conventional mathematical tools with operational data and physical models of the power system components. Such an analysis is valid under a given contingency and operating condition. It may not be computationally feasible to simulate all possible combinations of contingencies that could result in partial or complete collapse of the power network as well as events of various switching actions and continuous load variations. Hence, it is difficult to assess the outage of a portion or the entire grid following unforeseen operating conditions. Moreover, the expansion of the bulk power systems, their physical behaviour and interrelationship intensifies the problem of comprehensive vulnerability study. Hence, to reduce and simplify these problems, it is important to deepen and contribute to traditional analysis with novel tools of vulnerability assessment.

The motivation to assess the vulnerability of grid networks employing complex network analysis is because of cascading failures of power grids globally following contingencies and other reasons. The North American power grid was the first analysed power grid [3] where the vertices were randomly removed, and in decreasing order of their degrees (of those vertices) the connectivity loss was monitored. The connectivity loss among various areas of the grid resulted in loss of potential of substations to draw power from the generators. The depletion of generation nodes (substations) was not effective in changing the average degree (connectivity) of the grid due to high redundancy at the generation nodes (substations). As the power grid is delicate towards the depletion of transmission nodes, the elimination of a single transmission line can lead to connectivity loss. The analytical methods in complex network theory revealed that the connectivity loss is substantially higher following planned attacks, particularly if the degree of attack is higher for high betweenness transmission hubs. Crucitti et al. [4] made the first reference to European power grids and compared the structural properties of Italian, French and Spanish power grids by identifying the elements whose removal genuinely affected the structure of those graphs. Rosato et al. [5] implemented high-voltage vulnerability assessment of power grid networks in Italy, France and Spain to analyse the damage caused by controlled removal of links. Thus, vulnerability assessments using topological methods are useful in assessing the weakness of an electrical network as well as to design distinct actions to minimise topological weaknesses.

5.2 VULNERABILITY INDICES: TOOLS OF VULNERABILITY ASSESSMENT

Several indices like *global efficiency, electrical betweenness, and netability*, defined in Chapter 4 can be used to identify and rank critical nodes (buses) and links (lines) in a power network. Indices of vulnerability can include electrical centrality, electrical betweenness, vulnerability assessment using netability and global efficiency. While global efficiency and netability are both indices representing vulnerability of the system when any network component (bus or line) loss occurs, betweenness is a measure of criticality of each of these components in a network. The reactance of the lines act as the weights of the edges of the graph along the shortest electric path, which is the path whose sum of weights is the smallest among all possible paths between the nodes. The complex power flow through any line or tie in a complex grid is governed by the line parameters as well as by bus voltages and power angles. Evidently, the magnitude of power transfer via any transmission line from the sending bus to the receiving bus is inversely proportional to the reactance of the line when the bus voltages are at nominal p.u. values. In this chapter, the line reactance has been assumed to be the weight of the transmission lines and it is presumed that the magnitude of power flow increases when the line reactance is low.

Some researchers [6] have employed metrics, such as degree and degree distribution, characteristic path length, clustering coefficient and betweenness, to determine how the relative connection of distributed generators (DGs) influences the topological structure of the grid network. Some researchers utilised weighted graph indices and proposed advanced metrics based on the topology and condition of practical working operation of the power networks for the measure of topological characteristics of the power networks employing DGs. Authors of [7] performed similar assessment on the basis of specific distinct metrics where they introduced three vulnerability indices: *Structural Vulnerability Index* (SVI), *Contingency Vulnerability Index* (CVI) and *Operational Vulnerability Index* (OVI). These indices were exploited to measure structural vulnerability to identify the vulnerable elements in the power network and assess the state of operation of the network. In this chapter, a new metric called Grid Vulnerability Index (GVI), which is a combination of the SVI, CVI and OVI, is calculated to assess the structural change in a transmission system due to incorporation of DG.

5.2.1   GRID VULNERABILITY INDEX (GVI)

GVI is a new index that assesses the overall functioning capability of a grid when the grid is equipped with various DGs fulfilling the load demand. Because line reactance is governed by the length of the line, the concept of electrical distance is introduced as the performance parameters of the power grid. The GVI can be expressed as [8]

$$GVI = \frac{1}{N_G N_D} \sum_{i \in G} \sum_{j \in D} \frac{P_i}{L_j e^{Z_{ij}}} \tag{5.1}$$

where N_G and N_D denote number of generation buses and number of load buses, respectively. P_i and L_j are the active power generation and maximum load of buses, respectively. Here, the term Z_{ij} would be zero if the DG is directly connected to the load bus. On the other hand, for conventional generators supplying power through transmission lines, the value of Z_{ij} would be higher. In this case, the attribution of generation bus to load bus reduces with the increase in the value of Z_{ij}.

To simulate GVI for a power network, the following assumptions are considered:

a) For every alteration in the allocation of DG in the power system, N_G varies. It rises with the addition of the DGs to the load buses.

b) The quantity of load buses is assumed to be constant despite being included with DGs.

c) DGs are allocated to load buses only as they are troublesome to synchronise with traditional generation buses.

d) The capacity of DG has been assumed to be half the capacity of peak load at load bus. The highest DG capacity is 50 MW, although few load nodes have more half of peak load than this.

5.2.2 Efficiency-Based Vulnerability Index

In general, the rating of a distributed energy resource (DER)-based generator (i.e. DGs) is typically small. Such types of resources are installed at load nodes in remote or isolated geographical area. The load nodes that are distant from these generators are minimally facilitated. Moreover, the capability of a DG drastically reduces as the distance between it and a load node increases. Thus, the index efficiency does not apply in analysis of the network vulnerability when the network is equipped with DG connected at remote load nodes. With respect to this, a novel index (metric) with the assumption of the local supplying characteristic of DG is proposed:

$$e_i = \frac{1}{P_{D_j}} \frac{1}{N_g} \sum_{i \in V_G} \frac{P_{G_i}}{2^{d_{ij}-1}}$$ (5.2)

where e_i represents the efficiency of supplying power to a load node j. P_{D_j} is the active load demand of node j, P_{G_i} is the power capacity of generator node i and d_{ij} is the shortest path between generation node i and load node j. N_g is the number of DGs in the network. The bigger the value of e_i is, the more will be the power supplying efficiency of node i.

Consequently, the Global Average Power Supplying Efficiency Index can be expressed as [8]:

$$e_i = \frac{1}{N_g} \sum_{i \in V_G} e_i = \frac{1}{N_g} \frac{1}{N_D} \sum_{j \in V_D i \in V_G} \frac{P_{G_i}}{2^{d_{ij}-1}}$$ (5.3)

The exponential function in Equation (5.3) describes the characteristic of power supplying efficiency of the DG which decreases sharply with the increase in the transmission distance.

Following any decrement in the value of e, the vulnerability of the network can be assessed using the following expression:

$$e' = \left(1 - \frac{e}{e_0}\right) \times 100\%$$ (5.4)

Here, e_0 denotes the power supplying efficiency before the failure and e denotes the power supplying efficiency after the failure. From Equation (5.4), it is evident that the bigger e', more will be the drop in the functioning of the network, and the affected node (bus) or link (line) will be more prone towards vulnerability.

5.3 ASPECTS OF VULNERABILITY ANALYSIS OF POWER TRANSMISSION NETWORK

Identifying the vulnerable components in a power grid [9] is vital to the design and operation of a secure and stable system. One aspect of vulnerability analysis is to identify those transmission lines for which minor perturbations in their conductive properties lead to major disruptions to the grid, such as voltage drops, or the need for load shedding at demand nodes to restore feasible operation. Most studies focus on minimising the costs of load shedding and additional generation in the DC model or in lossless AC models. In case of identification of critical components of a power system, the grid has to be modelled with full AC power flow equations, which are the most accurate mathematical models of power flow.

A second aspect to measure severity of an attack is to consider the minimum adjustments to power that must be made to restore the grid to feasible operation. Power adjustments take the form of shedding load at demand nodes and adjusting generation at generator nodes. This is explained in the Chapter 6 in the islanding procedure for modelling cascading failure in power transmission grids where load redistribution model is followed using preferential probability. The next section discusses the application of vulnerability indices in power transmission network by simulation.

5.4 SIMULATION FOR VULNERABILITY INDICES IN POWER TRANSMISSION NETWORK

5.4.1 Using Global Efficiency and Netability Method

Figure 5.1 shows IEEE 14 bus system [10,11] whose vulnerability is being assessed by obtaining the global efficiency and netability values of the lines of this network.

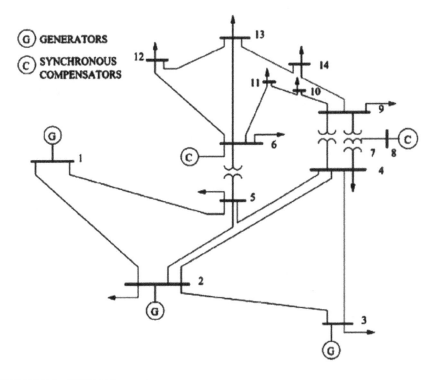

FIGURE 5.1   IEEE 14 bus system.

TABLE 5.1

Line Parameters of IEEE 14 Bus System

Line No.	Connecting Nodes	Line Reactance	Line Flow Limits
01	1–2	0.05917	130
02	1–5	0.22304	130
03	2–3	0.19797	65
04	2–4	0.17632	130
05	2–5	0.17388	130
06	3–4	0.17103	65
07	4–5	0.04211	90
08	4–7	0.20912	70
09	4–9	0.55618	130
10	5–6	0.25202	32
12	6–11	0.19890	65
12	6–12	0.25581	32
13	6–13	0.13027	65
14	7–8	0.17615	65
15	7–9	0.11001	65
16	9–10	0.08450	65
17	9–14	0.27038	32
18	10–11	0.19207	32
19	12–13	0.19988	32
20	13–14	0.34802	16

The line reactance is considered as weights of the edges of the graph. The line parameters that are mainly considered are shown in Table 5.1:

Sample Calculation:

Global efficiency (V_e): From Equations (4.14) and (4.15), the global efficiency can be calculated as follows:

For the given network

$$\sum_{i \neq j} \frac{1}{d_{ij}} = 46.31667, \quad E = 0.254487$$

For line no. 1, $\sum_{i \neq j} \frac{1}{d_{ij}} = 45.65, \quad E_1 = 0.250842$

Global efficiency for removal of line 1 is then given by

$$V_{e1} = \frac{E - E_1}{E} = 1.43\%$$

Netability (V_a): From Equations (4.23) and (4.24), the netability can be calculated.

In the same network,

$$\sum_{i=G}\sum_{j=D}\frac{C_{ij}}{Z_{ij}}=18218.0365, \quad A=506.0565$$

For line no. 1, $A_1 = 459.7372$

Netability for removal of line 1 is, thus, given by

$$V_{a1} = \frac{A - A_1}{A} = 9.15\%$$

Following the above method of calculation, the values of global efficiency and netability for the given network are obtained and shown in Table 5.2.

The line vulnerabilities are calculated using Equations (4.15) and (4.24) and the comparable graphs are resulted as shown in Figure 5.2:

It is observed that netability metric that includes the electrical properties of extended topological approach gives enhanced results of vulnerability analysis. The critical lines can be identified as listed below in Table 5.3.

TABLE 5.2
Calculated Values of Global Efficiency and Netability

Line No.	Connecting Nodes	V_e (%)	V_a (%)
01	1–2	1.43	09.15
02	1–5	2.73	08.44
03	2–3	2.15	17.57
04	2–4	3.6	14.14
05	2–5	2.08	05.47
06	3–4	3.63	19.41
07	4–5	8.44	18.96
08	4–7	8.5	17.17
09	4–9	1.76	11.58
10	5–6	8.2	01.73
12	6–11	9.31	18.86
12	6–12	4.06	05.24
13	6–13	7.12	09.14
14	**7–8**	**7.19**	**29.6**
15	7–9	8.25	17.8
16	9–10	0.6	16.45
17	9–14	4.1	08.93
18	10–11	4.46	14.03
19	12–13	1.43	05.02
20	13–14	5.75	12.04

Bold values are having the highest value, it is the most vulnerable line

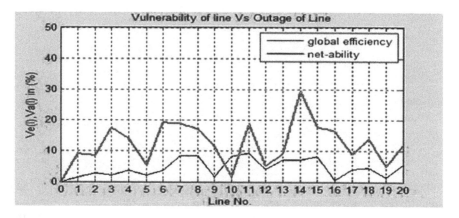

FIGURE 5.2 Graph of vulnerability analysis of lines using global efficiency versus netability of IEEE 14 bus power system.

TABLE 5.3
Critical Lines Identified from Graph

Line No.	Connecting Nodes	Vulnerability in (%)
3	2–3	17.57
6	3–4	19.41
7	4–5	18.96
8	4–7	17.17
11	6–11	18.86
14	7–8	29.60

From the above, it is noticed that line number 11 is highly vulnerable in the given network which connects a generation bus to a transformer node. One of critical lines, that is, line no. 3 connects two generation buses. Other critical lines connect transformers and buses. For obvious reasons, these lines are vulnerable as they are associated with critical components such as transformers and generation buses of the power grid network.

The grid network has been analysed separately using only topological model (global efficiency) and by adding some power grid characteristics to the topological model (netability). The IEEE 14 bus system is studied and the critical lines are identified which justifies the applicability of the developed model, that is, the netability model. The analysis identifies critical lines in the power grid which is very important in maintaining the network security.

5.4.2 Using Betweenness and Netability Metric

The given network has been altered to obtain bus impedance matrix directly. Bus number 1 is considered as a slack bus in case of electrical betweenness and is numbered as the 0th node. According to [12], the line parameters and figure are shown in Appendix. On the basis of pure topological betweenness or betweenness centrality (Equations 4.8 and 4.9) and extended electrical betweenness (Equations 4.10 and 4.11), the simulation has been conducted on the test system to find line betweenness and node betweenness with respect to respective line and node numbers (Figures 5.3 and 5.4a–d).

From the above analysis, it has been observed that the lines exhibit different betweenness among themselves. Figure 5.4a reveals that line 70 has the largest value of betweenness. Other lines like line number 42, 71, etc. follow line 70 in descending order of magnitude of betweenness. Because betweenness interprets the criticality of an element connected between two nodes, it is obvious that ranking lines in the order of criticality is possible using this method of analysis.

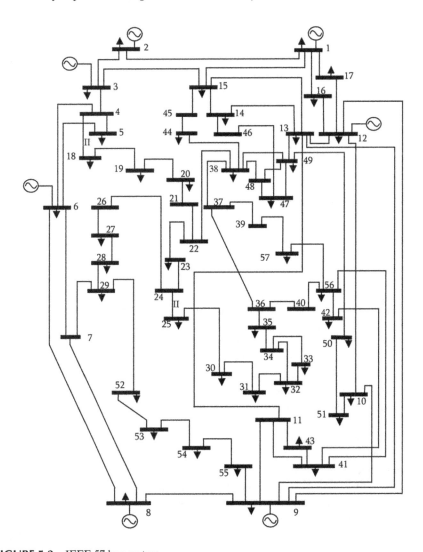

FIGURE 5.3   IEEE 57 bus system.

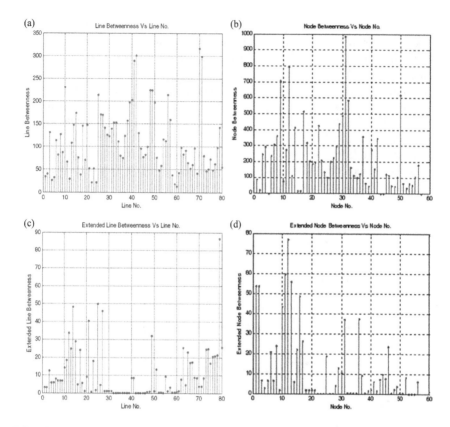

FIGURE 5.4 (a) Line betweenness of IEEE57 bus system, (b) extended line betweenness of IEEE57 bus system, (c) node betweenness of IEEE57 bus system and (d) extended node betweenness of IEEE57 bus system.

In the next step, the concept of extended betweenness has been considered. This extended betweenness indicates the criticality taking into account constraints like line flow limits, maximum transmission capability, and power transmission distribution factor. This concept is more realistic, and applying this concept in the test system, the simulation reveals that line number 79 showed the highest magnitude of extended betweenness followed by line number 25, 14, 27 and 21 (Figure 5.4b). Hence, the extended betweenness exhibits more specific criticality of the lines. The same exercise has been conducted for nodes (buses) revealing that bus number 12 exhibits the highest degree of extended betweenness, that is, criticality.

In the next phase, analysis and simulation has been conducted on the same test system to determine the netability of the system. The netability exhibits the degree of vulnerability of the power network following contingencies. While observing the results (shown in Figure 5.5), it is clear that despite line number 79 displaying high betweenness under normal operating conditions, failure of line number 14 indicates the drop of netability of 14.37% compared to other lines (case of single line outage). In this context, line number 14 has been taken into account as a specific case of

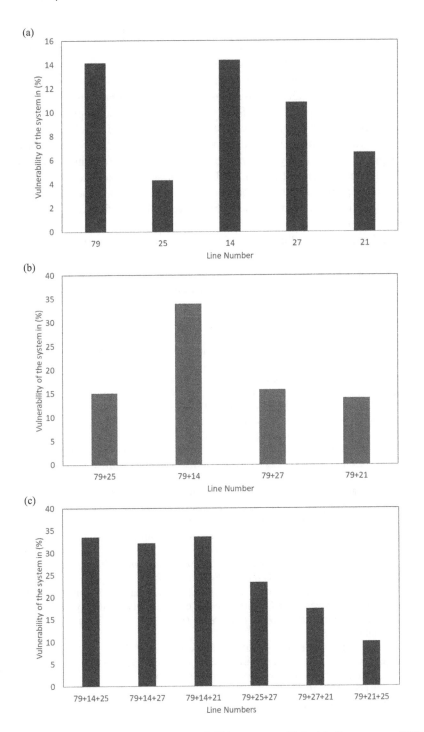

FIGURE 5.5 (a) Single-line outages of IEEE 57 bus system, (b) double-line outages of IEEE 57 bus system and (c) triple-line outages of IEEE 57 bus system.

interest. Thus, in multiple contingency studies, line number 14 is considered with other lines; as in case of two outages (line number 79 and 14) and in case of three outages (line number 79, 14 and 25). It reveals that the netability magnitudes remain high indicating higher vulnerability of the system line 14 is associated in contingency cases. Therefore, with undivided attention of operating personnel, the network can be arrested from catastrophic failure if line number 14 is checked on priority.

Hence, complex network theory can be successfully applied in vulnerability analysis of a sub-transmission system in conjunction with netability analysis. It is observed that computation of netability of the network facilitates ranking of critical lines and can be utilised in real power grid system. IEEE 57 bus system as a complex network is analysed in both topological and electrical context separately on the standard. The number of vulnerable lines has been obtained which reveals the practicality of the preferred model (i.e. the network modelled with extended betweenness). It may be suggested that topological analysis including electrical properties of the power network could serve as a complementary tool that can be used in decision-making by the operating personnel. Thus, application of complex network theory in power network can suitably be employed for identifying critical lines and buses in a sub-transmission system.

REFERENCES

1. L.H. Fink and K. Carlsen, "Operating under stress and strain", *IEEE Spectrum*, vol. 20, pp. 1357–1365, 2005.
2. http://en.wikipedia.org/wiki/Vulnerability_assessment.
3. R. Albert, I. Albert, and G.L. Nakarado, "Structural vulnerability of the North American power grid", *Physical Review E*, vol. 69, 2004.
4. P. Crucitti, V. Latora, and M. Marchiori, "Locating critical lines in high-voltage electrical power grids", *Fluctuation and Noise Letters*, vol. 5, pp. L201–L208, 2005.
5. V. Rosato, S. Bologna, and F. Tiriticco, "Topological properties of high-voltage electrical transmission networks", *Electric Power Systems Research*, vol. 77, pp. 99–105, 2007.
6. L. Feng, M. Yanbin, and M. Shengwei, "On the topological characteristics of power grids with distributed generation", *In the Proceedings of the 29th Chinese Control Conference*, Bejing, China, pp. 4714–4720, July 2010.
7. Z. Chen, C. Liu, Q. Xu, and C.L. Bak, "Vulnerability evaluation of power system integrated with large scale distributed generation based on complex network theory", *In Proceedings of the 47th IEEE International Universities Power Engineering Conference*, Middlesex, U.K, pp. 1–5, September 2012.
8. T. Chowdhury, A. Chakrabarti, and C.K. Chanda, "Analysis of vulnerability indices of power grid integrated DG units based on complex network theory", *In Proceedings of the 2015 Annual IEEE India Conference (INDICON)*, New Delhi, pp. 6540–6546, December 2015.
9. T. Kim, et al., "Vulnerability analysis of power systems", 2015.
10. http://www.ee.washington.edu/research/pstca/pf14/pg_tca14bus.htm.
11. H.B. Plittgen, "Computational cycle time evalution for steady state power flow calculations", December 1985.
12. http://www.ee.washington.edu/research/pstca/.

6 Analysis of Cascading Failure and Islanding in Grid Network

6.1 INTRODUCTION

In most countries power grid is a typical complex network that is distributed throughout a vast geographical region. Seldom, grid operation poses a threat of failure because of localised faults, consequently, affecting a part or the entire network leading to vulnerable collapse of the grid network. Such a failure can affect a large number of customers resulting in significant economic disruptions. Although most electrical failures of line(s) and or buses emerge and dissolve locally, some trigger avalanche mechanisms that can lead to failure of lines and buses as a chain event and eventually cripple the entire network [1–3]. Such vulnerable failure of an electric grid that leads to islanding of the grid network is termed as cascading failure. Literature review reveals that a number of power grids across the world have suffered cascading failures [1–5] causing blackouts. Cascading failure can achieve massive scale, with millions of customers affected by loss of billions of dollars [6,7].

In mathematics a popular model, called the "sand pile" [8] model, states that when sand is continuously piled in a heap, a point arises when a portion of the pile suddenly begins to subside. If an attempt is made to prevent the breakdown by adding more sand, the pile just collapses. Similarly, we can say that as a power grid approaches a critical point, the possibility of a collapse increases. With the addition of new elements to the grid, increasing demand, interconnections and increase in power flows, the grid may be quickly approaching its critical point. At such a point, it becomes necessary to study the complex dynamics arising in the grid and to find possible solutions to problems that may arise because of these dynamics.

> **Cascading failure [2]** is common in power transmission grids when one of the elements fails (completely or partially) and shifts its load to nearby elements in the system. Those nearby elements are then pushed beyond their capacity and they become overloaded and shift their load onto other elements.

The probable causes of cascading failure mainly involve:

- Severe overload
- Resource exhaustion
- Defective load management
- Inaccurate capacity planning
- Delay in communication response
- Sudden load burdening

The concerning issues associated with cascading failures are:

- They can take down the entire power system, tumbling the service of each component of the system one by one, until the entire load-balanced service is unhealthy.
- They are an exceptional type of failure from which it is hard to recover. They normally start with some small perturbation – like a transient network issue, a small spike in load or the failure of a few instances. Instead of recovering to a normal state over time, the system enters into a worse state. A system in cascading failure won't self-heal; it can only be restored through human intervention.
- If the right conditions exist in the system, cascading failures can strike with no warning. Unfortunately, the basic pre-conditions for cascading failures are difficult to avoid: it's simply failover. If failure of a component can cause load to shift to other parts of the system, then the basic conditions for cascading failure are present. However, there are patterns we can apply that help us defend our systems against cascading failures.

Among the above-stated probable causes of cascading failures, two scenarios are commonly observed [8] to be frequent. These are *load growth* and *random fluctuations* or perturbations.

Load growth is the increase in load in a specific ratio either intentionally or due to sudden high current operation.

Random perturbations can occur in the grid for several reasons. If random fluctuations are small, their effect may not be felt. However, if the fluctuations are considerable in magnitude, they can cause several undesirable effects on the functioning of the grid, including the initiation of a cascading failure. For example, random perturbations can occur due to the incorporation of renewable distributed generators, such as wind turbines and solar panels, in the grid.

The complexity of the entire power grid makes it difficult to model each and every individual component and study the stability of the entire system. Therefore, it is often the case that abstract models of the working of the power grid are constructed and analysed. The lack of a complex network-based model for the analysis of power grid was a major drawback of the studies conducted earlier. However, *Overload Cascade model* [9] was introduced to the complex network approach-based model. This model assigns capacity to the links in the network proportional to the initial power owing through them in the absence of any disturbance. When a disturbance occurs, these power flows are redistributed and the flow dynamics of the network are observed. The scenarios on which grid dynamics are observed include load growth and random fluctuations or perturbations.

In this model, the initial load and generation on the nodes represent the initial state of the system. Power flows are calculated using the DC Power Flow model. Every link has a capacity which determines the maximum amount of power that can be carried by that link. During normal operation, the system is stable and all power flows are within limit. If a disturbance, such as failure of an element, sudden increase in demand or load fluctuations, occurs in the system, all power flows are recalculated using equations and utilisation of all links is checked to see whether they are within their capacity. If any link exceeds its capacity, it is removed from the system, and its power is redistributed among other links, depending on impedances of the other links. Even though the model is simple, the presence of the flow equations guarantees that Kirchhoff's and Ohm's laws are properly considered, thus making the model more realistic for use with the power grid. The reduced complexity of this method allows its use even with large systems and provides a reasonable balance between complexity of the method and accuracy of the results.

In load growth scenario, the load growth is modelled by increasing all loads simultaneously step-by-step by a factor. Each time the loads are increased, the Overload Cascade model is run to calculate the power in the network. The increase in the demand causes increased power flows through the links. Because each link is bound by a finite capacity, some of the links exceed their capacity and get overloaded to supply the increased demand. These overloaded links fail and the power that was being carried through them is redistributed among the other links in the network. Because of this redistribution, more links may reach their capacity and fail. This may lead to a cascade of overload failures. This simulation can be performed with different values of the factor, and the fraction of links that failed at the end of each simulation can be recorded.

If a transmission line within the power grid trips, its power flow is automatically shifted to the neighbouring line(s), which in most of the cases are capable of operating with the extra load. However, in some cases, one or more of these neighbouring lines may get overloaded and exceed the allowable limit of power flow leading to overloading of line(s). This may lead to outage of the neighbouring overloaded line(s) which phenomenon may spread over other line(s) suffering from overloading. This process of failure of multiple lines as a chain event is a blow to the contingent power grid which is likely to collapse while encountering multiple line tripping. To better understand such events, the power grid needs to be analysed from a network perspective taking the advantage of concepts in complex network theory [10].

6.1.1   FUNDAMENTAL MODELLING APPROACH IN ISLANDING

A grid network can be modelled as a weighted graph having nodes or vertices (the buses) and links or edges (the transmission lines) [11]. When there is no direct transmission line from the node i to node j, the off-nominal entry (a_{ij}) of such a matrix is zero, otherwise a_{ij} is 1 provided a line exists between node i and node j. In a realistic approach of a power grid network model, each generator is assumed to transfer power to all buses (nodes) through transmission lines (links).

Following established concepts on complex network theory [12,13], the electrical load on a load bus can be defined as the electrical betweenness of that bus (node) associated with transmitting capabilities with number of most efficient paths passing from generators to load buses through the particular node. The same concept of electrical betweenness is valid for a transmission line where the line can be associated with an ultimate power transfer capacity C_{ij} that is proportional to its loading in the unperturbed state of the network.

Due to external causes when tripping of a line or bus connected to transmission line(s) occurs, the most efficient paths through which electrical power from each generator can reach each bus changes and the load gets redistributed among the network. Thus, a new value of electrical betweenness emerges for the surviving adjacent lines. The increment in load of the adjacent line(s) would raise the value of electrical betweenness for the line, and because the capacity C_{ij} of the line is characterised by the maximum load that the line can handle, the surviving adjacent line(s) would face outage provided the maximum value of the betweenness exceeds the ultimate capacity C_{ij} of the concerned line.

6.2   CONCEPT OF ISLANDING

Due to massive expansion of power system, large blocks of power are transmitted through long-distance EHV transmission lines from source to load centre in most power grids. Because of complex network structure, the grid suffers from possibility of lesser stability and is prone to encounter different types of contingencies and attacks that may lead to failure of a portion or the complete grid. This affects the capability of the power system to operate in a stable mode with high degree of reliability. In case the grid suffers from major contingencies and attacks, it may become vulnerable and there have been many occurrences of cascading failures in the recent past due to planned and unplanned outages of critical components of the grid. The cascading failure has a devastating effect on power system operation and the economy of the state. In western countries, researchers are assessing the possibility of cascading failure of the grid when the grid is subjected to electrical contingencies and unplanned outages. Defined by the North American Electric Reliability Corporation (NERC), a cascading failure is "the uncontrolled loss of any system facilities or load, whether because of thermal overload, voltage collapse, or loss of synchronism, except those occurring as a result of fault isolation" [14].

There have been suggestions such as moderating the dynamic equilibrium of the grid system to a point of self-organised criticality and strategic load shedding schemes so that it is possible to reduce the effects of cascading failure. Practical experiences reveal that intentional islanding of the power system is one such strategy. In intentional islanding, the utility through a pre-program software goes for the intentional splitting of the grid into separate controllable parts or islands. Each island must have its own independent generation, and the presence of distributed generation is preferred. In addition to intentional islanding, some strategic load shedding may be done to balance the generation and load in the sub-systems. Intentional islanding can be very helpful in isolating failures or localising them within the region where they occurred and preventing them from spreading throughout the system.

To ensure that the power grid survives following a major contingency or an attack on its critical parts employing intentional islanding is one of the solutions; though it is not always wise to opt for such a remedy until there is an utmost need for it. The utmost need is generated when the system suffers from cascading failure. The critical elements in the grid network consists of vulnerable buses and lines, and in case there is an outage of such component, the system may suffer tremendous damage and is prone to have cascading failure. If any of these vulnerable links happen to be the initial failure, islanding must be initiated. Using the vulnerability index, proposed in [15], it is possible to obtain the order in which the links fail.

6.3 MODELLING OF CASCADING FAILURE AND ISLANDING

Cascading failure in power transmission network is a potential threat which is usually triggered by a small and benign fault in a part of the transmission network and subsequently causes islanding of the power transmission system. In the transmission network model, a bus is treated as a node (or vertex) while a transmission line is treated as a link (or edge). An interconnecting path within a pair of nodes is defined as a sequence where any node is not repeated more than once. The geodesic distance is defined as the path of minimal length between a pair of nodes, that is, the shortest electrical distance between the pair of nodes. Obviously, the transmission of electrical power between a pair of nonadjacent buses depends on the transmission lines in the geodesic paths incorporating respective nodes (buses). The vulnerability of a bus (node) in the power transmission network can be assessed by the quantum of geodesic paths passing through it and is conventionally denoted by the *betweenness centrality* of that bus [16].

The pure topological approach of analysing a network can be employed to find critical elements of a transmission power network. However, this may not accurately address identifying vulnerable elements in a power grid as such a transmission grid possesses typical characteristics which needs to be taken into account while analysing the vulnerability of the elements in the grid. This leads to the concept of *electrical betweenness* [13] where power transmission network buses are categorised with respect to their characteristics, and each power line performs in power transmission system subjected to operating equality and inequality constraints.

6.4 APPLICATION OF PREFERENTIAL PROBABILITY IN MODELLING OF CASCADING FAILURES

Cascading failure in power transmission network is a potential threat which is usually triggered by a small and apparently benign fault in a part of the transmission network and subsequently causes islanding of the power transmission system An interconnecting path within a pair of nodes is defined as a sequence where any node is not repeated more than once. Obviously, the transmission of electrical power between a pair of nonadjacent buses depends on the transmission lines in the geodesic paths incorporating respective nodes (buses). The vulnerability of a bus (node) in the power transmission network can be assessed by the quantum of geodesic paths passing through it and is conventionally denoted by the *betweenness centrality* of that bus [16].

The pure topological approach of analysing a network can be employed to find critical elements of a transmission power network. However, this may not accurately address the identification of vulnerable elements in a power grid as such a transmission grid possesses typical characteristics which needs to be taken into account while analysing the vulnerability of the elements in the grid. This leads to the concept of *electrical betweenness* [13] where power transmission network buses are categorised with respect to their characteristics and each power line performs in power transmission system subjected to operating equality and inequality constraints.

As stated earlier, betweenness of a bus (or node m) in a power network can be defined as [13].

$$T(m) = \frac{1}{2} \sum_{g \in G} \sum_{d \in D}^{N_G \, N_D} C_g^d \sum_{l \in L^m} \left| f_l^{gd} \right| \tag{6.1}$$

where $m \neq g \neq d \in L^m$

$\sum_{l \in L^m} \left| f_l^{gd} \right|$ represents the summation of PTDF of all the lines connecting bus m, power

being injected at bus g with power withdrawal at bus d.

$\frac{1}{2} C_g^d \sum_{l \in L^m} \left| f_l^{gd} \right|$ represents the quantum of transmitted power available at bus m during transmission of electrical power from bus g to bus d. G represents the set of generation buses; D denotes the set of load buses; N_G and N_D represent the number of generation buses and the number of load buses, respectively; L^m is the set of lines connected with bus m.

Electrical betweenness of a transmission line (link l) in a power network model can then be represented as [13]

$$T(l) = \max \left[T^p(l), \left| T^n(l) \right| \right] l \in L \tag{6.2}$$

where $T^p(l)$ and $T^n(l)$ are positive and negative electrical line betweenness of the transmission line l.

$$T^p(l) = \sum_{g \in G} \sum_{d \in D} C_g^d f_l^{gd} \qquad g \neq d, f_l^{gd} > 0 \tag{6.3}$$

$$T^n(l) = \sum_{g \in G} \sum_{d \in D} C_g^d f_l^{gd} \qquad g \neq d, f_l^{gd} < 0 \tag{6.4}$$

Here, $C_g^d f_l^{gd}$ represents the power flow in the line l. The set of electrical betweenness qualifies the contribution of the grid network component, for example, a transmission line, to the power grid network, and thus the criticality of elements of the grid network can be assessed. With higher electrical betweenness, the criticality of the line is enhanced. Each line can then be associated with a definite value of electrical betweenness $T_{gd, \max}$ where

$$T_{gd,\max} = \varepsilon * T(l) \tag{6.5}$$

Here, ε is the tolerance factor that represents the ability of a line to handle increased electrical betweenness to retain the line operation within specified limits. It may be noted that $\varepsilon > 1$. A transmission line may be termed as overloaded, making it vulnerable to failure, if $T_{gd, max} > C_{gd}$.

When considering power transmission [17] from a generator bus g to a load bus d, as PTDF has a sign, if we specify a reference direction for line l the PTDF of f_l^{gd} should be positive, negative or zero. Then, we determine positive betweenness $(f_l^{gd} > 0)$ and negative betweenness $(f_l^{gd} < 0)$ by Equations (6.3) and (6.4), respectively. In Equation (6.3), if there is no $f_l^{gd} > 0$, then $T^p(l) = 0$ and in Equation (6.4), if there is no $f_l^{gd} < 0$, then $T^n(l) = 0$.

The outage of a line or a bus connected to a line alters the shortest paths between the buses (nodes), and consequently, the power flow of the tripped line gets redistributed following the concept of preferential probability Π_d. Such a process may create overloads and subsequent overloading in other adjacent lines. The electrical betweenness $T_{gd, max}$ of the adjacent line(s) may exceed the corresponding limit(s) C_{gd} for the overloaded line(s) following load redistribution and may trigger an avalanche chain mechanism collapsing the entire system. During load redistribution, the preferential probability can be expressed as [18]

$$\Pi_d = \frac{c_d^\alpha}{\sum_{m \in N} c_m^\alpha} \tag{6.6}$$

where α is a tunable parameter that can be varied from 0.2 to 2.2 in small-world networks [9,19–23] like power grid, c is the degree (or connectivity) of d and N is the set of neighbours of d.

Obviously, following collapse of a overloaded line, the additional electrical betweenness ΔT_{gd} transferred to an adjacent line is then given by

$$\Delta T_{gd} = T(l)\Pi_d \tag{6.7}$$

This adjacent line may face outage and would induce further redistribution of power flow making other surviving lines potentially vulnerable for collapse provided for any surviving line, $T_{gd, max} > C_{gd}$.

$$\text{Also, } T_{gd,max} = T(l) + \Delta T_{gd} = \varepsilon * T(l) \tag{6.8}$$

This iterative process for removal of an overloaded line followed by redistribution of loads among adjacent lines with subsequent identification and tripping of adjacent line(s) with high degree of betweenness leads to cascading failure resulting in islanding condition.

6.5   SIMULATION

To illustrate the dynamic behaviour of cascading failures, the developed concept has been validated first in IEEE 57 bus test system [24] and then in the part of the Eastern grid system of India (203 bus, 267 lines and 24 generator electrical grid network). The single line diagram of both the test systems is shown in Appendix A and B respectively.

The following algorithm illustrates the process of testing a grid system against possible cascading collapse following tripping of a line with highest electrical betweenness.

Step 1: A generation dispatch process is performed with generation-load balance using Optimal Power Flow program employing Newton–Raphson method in grid systems under consideration. The output exhibits parameters of grid operation including power flow in the lines.

Step 2: Loading at heavily loaded buses are increased and generation-load balance is re-established. The network power flow model is executed to calculate subsequent power flow in each line.

Step 3: Electrical betweenness of the lines (links)/buses (nodes) for the model network are calculated.

Step 4: The line (link) having the highest value of electrical betweenness is considered to be the overloaded line. The bus (node) with the highest electrical betweenness is assumed to be a potential threat facing bus failure.

Step 5: The overloaded line is tripped and/or the bus with highest electrical betweenness is removed following a simulated failure. The loading of adjacent lines is calculated using the developed concept of power flow redistribution associated with the concept of preferential probability for load redistribution.

Step 6: Criterion of overload (i.e. $T_{gd,\,max}$ (the sum of $T(l)$ and $\Delta T_{gd}) > C_{gd}$) for each of the adjacent lines is checked.

Step 7: Any line which satisfies the criterion $T_{gd,max} > C_{gd}$ is overloaded and tripped with subsequent redistribution of power flow to adjacent lines.

Step 8: Computations in Steps 6 and 7 are executed for these adjacent lines. If this process ends with surviving lines having $T_{gd,max} \leq C_{gd}$, the system survives. However, if this process ends with an ultimate line where $T_{gd,max} > C_{gd}$, the system is subjected to cascading failure and islanding may occur.

In the simulation, the IEEE 57 bus test system has first been tested for failure of highest betweenness line as well as for failure of the bus with highest electrical betweenness. In this process, the transmission line with the highest electrical betweenness in the IEEE 57 bus test system has been identified (line 79) and removed in the simulation. This induces load redistribution among adjacent lines (line 10, 11, 13, 15 and 25). Line 11 has now the highest electrical betweenness and is removed (Table 6.1). The simulation reveals that continuation of the iterative process for removal of an overloaded line followed by redistribution of loads among adjacent lines with subsequent identification and tripping of adjacent line(s) with high degree of betweenness leads to islanding condition. It is clear from Table 6.1 that at the end of the process only line 21 survives, and tripping of the this line leads to termination of the process and the system is islanded. The mapping diagram of this process is shown in Figure 6.1 (circles indicate the islanded nodes while the rectangles are the line numbers). This diagram depicts the mapping pattern of cascading failure, and it may be noted that the tripping or cascading failure occurs only in the direction shown by arrows and not otherwise or in any reverse process.

TABLE 6.1

Results on Tripping of Line 79

Line No.	Adjacent Lines	π_d	$T_{gd,max}$	C_{gd}
79	10	0.4836	22.8165	70
	11	0.3631	25.8475	20
	13	0.2362	31.8237	30
	15	0.5352	45.4841	50
	25	0.4849	74.8006	80
11	10	0.4351	22.0706	70
	12	0.2672	43.0848	30
	13	0.3962	35.9426	30
	15	0.2436	36.8447	50
	74	0.5964	41.2983	45
12	21	0.5632	64.6761	50
	23	0.2975	24.6177	25
	13	0.1326	29.1567	30
	74	0.8973	49.0825	45
21	22	0.3261	1.6372	50
	23	0.8964	35.9826	25

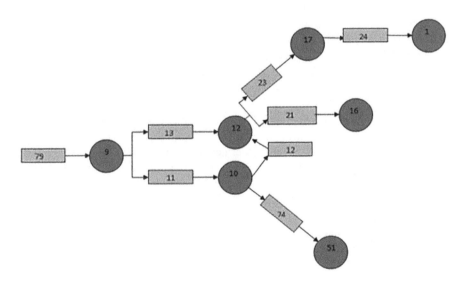

FIGURE 6.1 Load redistribution resulting islanding initiated by outage of line 79.

In the next step, standard load flow program is executed applying contingency with tripping of lines, as indicated in the preceding steps. It has been found that load flow convergence is possible till line 23 is tripped. Following tripping of line 23, the load flow does not converge, indicating that the system is subjected to failure.

In the next step, the bus with the highest electrical betweenness (node 12) is tripped in IEEE 57 bus test system. This induces load redistribution among adjacent

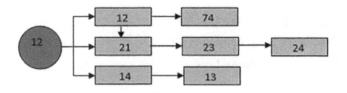

FIGURE 6.2 Load redistribution resulting islanding initiated by outage of node 12.

lines (line numbers 12, 13, 14, 21 and 23). The simulation reveals that continuation of the iterative process for removal of an overloaded line followed by redistribution of loads and subsequent identification and tripping of line(s) with high degree of betweenness leads to cascading failure. The mapping pattern of this process is shown in Figure 6.2.

In the 203-bus test system, the same process is repeated, and it has been observed that cascading failure process is initiated with removal of a single line with the highest electrical betweenness (line number 206). Outage of line 206 results in redistribution of power flow associated with preferential probability in the neighbouring lines, as shown in Table 6.2.

It reveals that for line number 14 and 13, the condition $T_{gd,\,max} > C_{gd}$ is satisfied. From outage of line number13, it can be observed from Table 6.2 that the criterion of overload appears in line number 165 and 17 (as for each of these two lines $T_{gd,\,max} > C_{ij}$ is satisfied). Following outage of line 165, the process of further redistribution of loads is terminated because it is one of the terminal lines creating the state of islanding for the 203 bus network. The results obtained in Table 6.2 have been pictorially depicted in Figure 6.3 where the rectangles represent the transmission lines (links or edges of complex network) and the circles represent the buses (nodes or vertices of complex network).

The process of cascading failure for the 203 bus network has been shown in Figure 6.4 following tripping of the node (node number 101) that has the highest electrical betweenness in the model power network. During islanding operation, the system parts into two or more sections and each section may tend to have overvoltage

TABLE 6.2
Results on Tripping of Line 206

Line No.	Adjacent Lines	π_d	$T_{gd,max}$	C_{gd}
206	207	0.4982	54.6062	85
	14	0.5277	48.9331	40
	13	0.7782	88.5479	80
	241	0.6962	65.5667	70
13	237	0.4932	48.8378	90
	165	0.6742	57.8635	45
	17	0.4642	62.9876	50
165	157	0.6329	96.5728	90

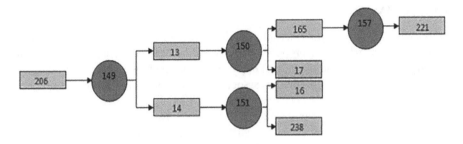

FIGURE 6.3 Load redistribution resulting islanding initiated with outage of line 206.

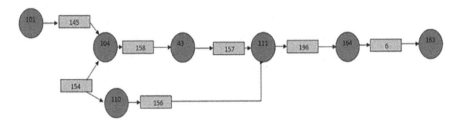

FIGURE 6.4 Load redistribution resulting islanding initiated by outage of node 101.

and undervoltage problems depending upon the amount of generation available in each section. The loading level, nominal rating, short-term emergency rating and long-term emergency rating for some of the network components (generators, transformers and lines) might have been violated under such a condition.

To check the results obtained, load flow program is again executed, and it has been observed that with sequence of tripping of lines as considered in the preceding portion leads to islanding of the 203 bus grid network, validating the use of electrical betweenness for assessing the cascading failure of a grid network.

The electrical betweenness of the transmission lines in the IEEE 57 bus test grid as well as the part of Eastern grid of Indian power system have been calculated following the concepts of complex network theory. The lines are arranged in descending order from the highest value of line electrical betweenness down to the lowest value. Tripping of a line, with highest betweenness value, has been considered in this simulation. Its load gets redistributed to the adjacent line(s) following preferential probability of load redistribution subsequent to outage of the line. The new values of line betweenness of the surviving adjacent lines are checked to identify the next overloaded line(s) prone to failure. Once such an overloaded line collapses, its load is again redistributed to other lines following preferential load redistribution, giving rise to possibility of further overloading in other surviving line(s) and subsequent tripping of such line(s). In this investigation, the roadmap of cascading failure with islanding condition of a grid network followed by failure of overloaded transmission lines in steps has been mapped to assess the vulnerability of the grid network under consideration.

6.6   SUMMARY

In this chapter, a topological cascading failure assessment approach is described for complex power grids. The property of electrical betweenness has been utilised to identify critical lines and tripping of such a line with the highest electrical betweenness resulting in load redistribution associated with preferential probability of load attachment. To validate the developed concepts, IEEE 57 bus test system and a typical 203 bus grid network of eastern India have been utilised. Outage of a line with the highest electrical betweenness within such a grid network has been simulated, and it has been observed that the redistribution of loads to the adjacent lines associated with preferential attachment leads to overload of an adjacent line(s). The iteration of simulated outage of such an overload line and determination of subsequent loadings on surviving adjacent lines is continued in this investigation so that the process is converged to a single overloaded line. It has been observed that tripping of this terminating line would result in islanding of the power grid. Hence, it is concluded that this method is an efficient technique for analysing and assessing cascading failure following outage of overloaded line(s).

REFERENCES

1. R. Albert, I. Albert, and G.L. Nakarado, "Structural vulnerability of the North American power grid", *Physical Review E*, vol. 69, p. 025103, 2004.
2. P. Hines, K. Balasubramaniam, and E.C. Sanchez, "Cascading failures in Power grids", *IEEE Potentials*, vol. 28, no. 5, pp. 24–30, 2009.
3. Yang Nan, Liu Wenying and Guo Wei, "Study on scale-free characteristic on propagation of cascading failures in power grid", IEEE 2011 EnergyTech, Cleveland, OH, pp. 1–5, 2011, doi: 10.1109/EnergyTech.2011.5948519.
4. L.L. Lai, H.T. Zhang, C.S. Lai, F.Y. Xu, and S. Mishra, "Investigation on July 2012 Indian blackout", *2013 International Conference on Machine Learning and Cybernetics*, 14–17 July 2013. doi: 10.1109/ICMLC.2013.6890450.
5. P. Crucitti, V. Latora, and M. Marchiori, "A topological analysis of the Italian electric power grid", *Physica A*, vol. 338, pp. 92–97, 2004.
6. R. Kinney, P. Crucitti, R. Albert, and V. Latora, "Modelling cascading failures in North American power grid", *European Physical Journal B*, vol. 46, pp. 101–107, 2005.
7. IEEE Working Group on Understanding, Prediction, Mitigation and Restoration of Cascading Failures, "Benchmarking and validation of cascading failure analysis tools", *IEEE Transactions on Power Systems*, vol. 31, no. 6, pp. 4887–4900, 2016.
8. S. Pahwa, "Dynamics on complex networks with application to power grids", Dissertation Thesis, Kansas State University, 2013.
9. A.E. Motter and Y.C. Lai., "Cascade-based attacks on complex networks", *Physical Review E*, vol. 66, p. 065102, 2002.
10. G.A. Pagani and M. Aiello, "The power grid as a complex network: A survey", *Physica A*, vol. 392, pp. 2688–2700, 2013.
11. K. Sun, "Complex networks theory: A new method of research in power grid", *In Proceedings of IEEE PES Transmission and Distribution Conference and Exhibition*, Asia and Pacific, Dalian, China, pp. 1–6, 2005.
12. D. Wu, E. Bompard, and F. Xue, "The Concept of Betweenness in the Analysis of Power Grid Vulnerability," In *Engineering. Complexity in*, Rome, Italy, pp. 52–54, 2010, doi: 10.1109/COMPENG.2010.10

13. E. Bompard, D. Wu, and F. Xue, "Structural vulnerability of power systems: A topological approach", *Electric Power Systems Research*, vol. 81, no. 7, pp. 1334–1340, 2011.

14. North American Electric Reliability Corporation, "Evaluation of criteria, methods, and practices used for system design, planning, and analysis response to NERC blackout recommendation 13c, 2005. http://www.nerc.com/docs/pc/tis/.

15. M. Youssef, C. Scoglio, and S. Pahwa, "Robustness measure for power grids with respect to casacading failures, *In Proceedings of the International Workshop on Modeling, Analysis and Control of Complex Networks*, San Francisco, CA, pp. 45–49, September 2011.

16. G. Rout, T. Chowdhury, and C.K. Chanda, "Betweenness as a tool of vulnerability analysis of power system", *Journal of Institution of Engineers (India): Series B*, Springer, vol. 26, no. 8, pp 2443–2451, 2016.

17. A.V. Gheorghe, M. Masera, and P.F. Katina, *"Infranomics: Sustainability, Engineering Design and Governance"*, New York: Springer Ltd.

18. J. Wang, L. Rong, and L. Zhang, "A new cascading model on scale free network with tunable parameter", *First International Conference on Intelligent Networks and Intelligent Systems*, IEEE Computer Society, Washington, DC, 2008.

19. X.F. Wang and G.R. Chen, "Complex networks: Small-world, scale-free and beyond", *IEEE Circuits and Systems Magazine*, vol. 3, no. 1, pp. 6–20, 2003.

20. P. Crucitti, V. Latora, and M. Marchiori, "Model for cascading failures in complexnetworks", *Physical Review E*, vol. 69, p. 045104, 2004.

21. K.-I. Goh, B. Kahng, and D. Kim, "Universal behaviour of load distribution in scale free networks", *Physical Review Letters*, vol. 87, no. 27, p. 278701, 2001.

22. R. Cohen, "Breakdown of the Internet under international attack", *Physical Review Letters*, vol. 86, no. 16, p. 3682, 2001.

23. Y. Zhang, S. Cai, C. Chen, and J. Shi, "Robustness of deterministic hierarchical networks against cascading failures", *IEEE* International Conference on Electrical and Control Engineering, USA pp. 4063–4066, 2011.

24. http://www.ee.washington.edu/research/pstca/.

7 Assessment of Resilience in Power Transmission Network

7.1 INTRODUCTION

The word resiliency is derived from the Latin word "resilio," which literally means the ability of an entity to rapidly get back to its original state, shape, health or position, after enduring and opposing stress applied on it. Based on these, resilience of networks, systems, infrastructure, machines, human health, community response to epidemics or natural disasters have been defined. Power grid resiliency is gaining importance as climate change increases the threats to modern infrastructure.

7.2 CONCEPTS OF RESILIENCY, ROBUSTNESS, RELIABILITY AND STABILITY

The concept of resilience is still in the early stages of research, and though there are multiple definitions, there is no universally accepted one. Overlying concepts of risk assessment, reliability, recovery and robustness contribute to definitions of resiliency. Some sources [1] reveal resilience of a system depends on how capable it is to decrease the intensity and extent of impact caused by the disturbances, whereas some researchers [2,3] define resilience as the ability of a system to identify risks.

Analysis of system performance focuses on system failure defined as any output value in violation of a performance threshold (such as a performance standard). System performance can be described from three different viewpoints: (i) how often the system fails (reliability), (ii) how quickly the system returns to a satisfactory state once a failure has occurred (resiliency) and (iii) how significantly strong the system remains to face the consequence of failure (robustness).

On the basis of power system operating state:

Robustness to a class of disturbances is defined as the ability of a system to maintain its function (normal state) when it is subjected to disturbances of this class.

Resilience to a class of unexpected extreme disturbances is defined as the ability of a system to gracefully degrade and to quickly self-recover to a normal state.

Reliability is the ability of the power system to deliver electricity to customers with acceptable quality and in the amount desired while maintaining grid functionality even when failures occur.

Or reliability is the probability that no failure occurs within a fixed period of time, often taken to be the planning period.

Resiliency will describe how quickly a system is likely to recover or bounce back from failure once failure has occurred. The resilience of a system presented with an unexpected set of disturbances is the system's ability to reduce the magnitude and duration of the disruption. A resilient system downgrades its functionality and alters its structure in an agile manner. If failures are prolonged events and system recovery is slow, this may have serious implications for the transmission system.

Robustness is the ability of a system to cope with a given set of disturbances and maintain its functionality. Robustness is concerned with strength, whereas resilience is concerned with flexibility.

Robustness and resilience belong to two different design philosophies. When a robust grid is attacked, it may break like an oak tree in a storm. When a resilient grid is attacked, it can bend and survive like a reed in a storm. From a system engineering point of view, absolute robustness can actually lead to fragility. The interrelationship between these three parameters can be explained by Figure 7.1.

In the wake of unprecedented disasters and attacks, robustness and resilience have become buzz words in many disciplines including biology, ecology, sociology, systems engineering and infrastructure engineering. The traditional definition of resilience in systems engineering is the capacity for fast recovery after stress and for enduring greater stress [4,5]. In systems engineering, resilience includes maintaining system functionality following disturbances. Robustness, on the other hand, explains the ability of a system to resist change without losing stability [6]. In some disciplines like social systems and organisational systems, the term resilience is similar to the term robustness. In infrastructural systems, and especially in power systems, however, the term robustness and resilience are more distinct; this is due to power systems structure and function centring on conductor lines delivering electric power to a certain area within specific voltage and frequency ranges.

Extreme robustness actually leads to fragility. Power systems are usually robust enough to withstand one contingency $[N - 1]$ or two contingency $[N - 2]$ events, where N stands for the number of system buses. However, beyond that, they are generally vulnerable. Moreover, the robustness is usually used with specific assumptions for protection system operation under pre-defined operational ranges for voltage and loading.

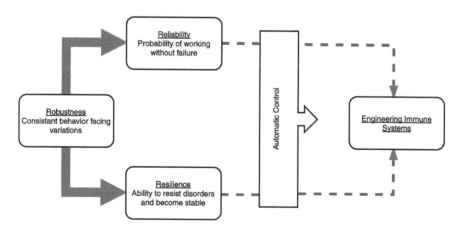

FIGURE 7.1 Block diagram of reliability, robustness and resilience in control operation.

The concept of reliability is also used in industrial and systems engineering and is accompanied by statistical and probabilistic approaches that characterise system performance after predicted and unpredicted failures. Reliability and stability are two more explicit power system concepts that pre-date the terms "robustness" and "resilience" [6]. Reliability and stability are well studied concepts in power systems. Similarities and dissimilarities between them and the terms robustness and resilience can inform future cyber physical resilience studies.

Discussions of resilience often centre around a system survivability that leverages load shedding, generation outages and other actions. Reliability is a measure of the system's ability to serve all loads. The system's ability to serve loads is traditionally referred to as service availability, which falls under the power systems definition of reliability. Reliability indices are usually expressed in terms of the probability of load loss [6]. The loss of load probability is expressed in days per year. Reliability is primarily concerned with the risk of service interruption of device failure.

"Stability is the ability of a system to remain intact after being subjected to small perturbations" [7]. In power systems, stability for a given initial operating condition means the system will regain operation equilibrium state after small perturbations. Stability is focused on the system equilibrium point. However, the concept of robustness in power systems goes beyond stability. To be robust, the electric grid has to be stable in the face of small perturbations as well as major equipment failures, man-made attacks and natural disasters [7].

Power system resiliency generally refers to the ability of the system to deal with unpredictability of catastrophic weather events and robustness of the infrastructure. Resiliency of transmission lines are [8–10] different from that of distribution lines. Transmission lines undergo extensive planning, maintenance, making them strong enough to withstand forces of nature. Distribution systems, on the other hand, have financial constraints for robust planning and 90% of the power outages are due to the inability of the distribution system network to withstand strong disturbances leading to power outages downstream of sub-transmission networks.

Metrics of resiliency are necessary to assess about preparation for ensuring proper operation and safety of existing electrical network when the network is subjected to contingency or attack. Resiliency implies the immunity of the electrical network and it is a subjective concept. It can be assessed through its metric but there is practically no way to quantify the resiliency. The resiliency related to the planning and operation of electrical power system deals with the effect of sudden disturbances to the network [11]. For long-term resiliency, it is important to determine how the power system infrastructure of a country would sustain itself in case imported raw materials suddenly become unavailable.

There are considerable amount of research work on system restoration and optimal recovery strategies in power system networks following a vulnerable disturbance. Usually, the system operators heavily depend on switching of different portions of the network to ensure proper restoration of the electrical network subjected to unplanned contingency or attack but in case the switches face catastrophic damages due to the vulnerability of the contingency or attack, then the restoration policy fails.

To assess the resiliency, due consideration may be given in the following aspects:

- A way of increasing the resiliency of the system is using microgrids. Microgrids are normally having self-sustaining resources. They are effective in several smart grid projects to prevent wide spread outages. When the multiple microgrid system islands from the grid, it may prevent cascading outages in the wide area network by lowering load demand. However, it does not guarantee that the critical loads within the micro grid would continue to operate properly. Metrics of resilience of the power network ensures the level of preparedness against unforeseen attacks with integration of multiple microgrids.
- Resiliency indices of microgrids can adjust with microgrid controllers or changes in design if it is possible to demonstrate remarkable enhancement in resiliency after implementing new algorithm or installation of new devices.
- In any electrical system it is very important that there is remarkable resiliency to ensure the robustness of critical electrical loads associated with military bases, hospitals and police stations.
- There is a possibility that resiliency have an impact in the economic aspects of power exchange between two microgrids in long term, or between a transmission company and a microgrid.
- Proper resiliency helps to improve the performance and design of the electrical protection scheme of the system, machinery and algorithms of operation.

Several isolated yet similar occurrences over the past emphasise the range of challenges power systems need to overcome. In addition, an eye must be kept on how these challenges are evolving due to climate change and emergence of new technologies. Strong winds, especially when combined with rain, hail and snow from seasonal storms, can damage electricity utility systems, resulting in service interpretations to large numbers of electricity customers, which is likely to include critical loads like airports, hospitals, city halls and other buildings deemed important to the community [12]. While most such power outages are caused by faults as a result of trees (or their branches) falling on local electricity lines and poles, major power outages tend to be caused by extreme weather events. Here are some of the examples:

- North-eastern states of the United States were hit by Hurricane Sandy in 2012, which destroyed over 1,00,000 primary electrical wires along with several substation transformers exploded and flooding numerous substations. These lead to the interruption of power supply, affecting ~7,000,000 consumers [13].
- During the summer of 2010–2011, Queensland, Australia was affected by widespread flooding that resulted in significant damage to six zonal substations and a large number of nodes, transformers and overhead wires. Approximately 150,000 consumers were affected by regular, long power outages over the entire two summers [14].

- In 2008, China had to bear the consequences of a severe ice storm, which led to failure of about 2000 substations, resulting in the collapse of about 8,500 poles, resulting long power disruptions in 13 provinces and more than 170 cities and towns across the world's most populated country [15].

Commonly used practices for resiliency are making lines underground, and/or removing the trees which caused faults, and/or taking mitigating action against rodents causing short circuit fault; however, these approaches are neither environment friendly nor a sustainable, economic and universal solution.

There is a huge drive to improve the resilience of the power infrastructure to such increasing occurrences of power outages due to abrupt weather disturbances. However, without having a formal procedure to measure the existing resiliency of the system, it may be an uncertain path to follow while reconfiguring the existing network to be more robust. Now, before proceeding to the details about assessment of resiliency, it is important to review the existing definitions and interpretations of resiliency.

7.2.1 DEFINITIONS OF RESILIENCY

The first formal definition of resiliency was published in 1973 by Holling [16], who said that resiliency is a form of persistence, the ability to absorb shock and changes and "still maintain the same relationships between populations or state variables." Gunderson et al. [17] modified the definition by adding buffer capacity for absorbing perturbations in a timely manner. Walker et al. [18] extended the definition to include the ability to self heal during disturbances. Kendra and Wachtendorf [19] described "bouncing back from a disturbance" as a crucial aspect of resilience. The breadth and number of definitions for "resilience" has increased significantly over the last decade, making it difficult to find a universal understanding of the term "resilience."

Merriam Webster defines resilience as "the ability to recover from or adjust easily to misfortune or change." There are numerous definitions of "resiliency" that have been introduced for any physical network infrastructure and the socio-economic system in the past decades [20–24]. In general, it can be referred as the ability of a system or an organisation to react and recover from unpredictable disturbances and events [25]. Zio [26] modified the concept of resiliency by stating "...systems should not only be made reliable, i.e. with acceptably low failure probability, but also resilient, i.e., with the ability to recover from disruptions of the nominal operating conditions." According to Haimes, "resilience is the ability of the system to withstand a major disruption within acceptable degradation parameters and to recover within an acceptable time and composite costs and risks." In 2013, White House declared "resilience is the ability to prepare for and adapt to changing conditions and withstand and recover rapidly from disruptions... [It] includes the ability to withstand and recover from deliberate attacks, accidents or naturally occurring threats or incidents." Committee on Increasing National Resilience to Hazards and Disasters, Committee on Science, Engineering, and Public Policy and the National Academies on their collaborative effort defined in 2012 that "resilience is the ability to prepare and plan for, absorb, recover from, and more successfully adapt to adverse events."

Though there are some distinctions existing among various definitions, it is worthy to compare them to reveal some relevant aspects of system operations.

Basically, resiliency addresses the concern that assesses the level of preparedness of a system encountering disruptions. It is important to clarify in resiliency assessment that how much of the service has been degraded and how quickly the service has been restored. Moreover, it is important to assess how completely the service has been restored. Resiliency also describes the degree of disruption across multiple dimensions, which could include type, quality, time and geography of service provision.

It is important to assess the state of the system from the viewpoint of its design and redundancy. For example, an electricity grid system that is designed with more redundancy, operated with more contingencies for back up and designed with recovery might experience a lesser and briefer disruption; and, if so, would be more resilient than a system that has less redundancy has fewer backups and is more difficult to rebuild.

Moreover, it may be noted that different responses lead to different resiliency at different costs. For example, with additional resources, it is possible to restore the electrical system while it is subjected to outages following disruptions. It is also important to utilise more efficient equipment, and, consequently, the quality of service provided after recovery could exceed the original level of service provided.

It is interesting to note that resiliency also depends on the timescale. With efficient maintenance an electrical network would suffer lesser disruption and even if there are disruptions, the restoration will be more prompt and quick. Indirectly, this would increase the resiliency of the power network.

Because resiliency is a subjective concept, definition of resiliency includes some additional relevant characteristics of the network. These characteristics are clarified as redundancy of the network, reliability, sustainability, vulnerability, fault tolerance capability and fast clearing circuit breakers, recoverability, etc.

7.2.2 Framework of Resiliency

The literature on resiliency shows that there are many different opinions about the "phenomenology" of resilience – which characterises resilient performance [24]. Therefore, instead of considering what resilient performance is, the proposed work considers what enables resilient performance, what makes it possible, and conversely what would make it impossible, if it was missing. This explains the philosophy of the motivation of the proposed framework. The proposed framework considered the diverse aspect of power systems' disruption and functionality. The first block diagram events set the base for the framework. Other blocks of the framework explain the characteristics of the grid on the basis of chosen events. Because power grid has diverse and continuous challenges as explained previously, it need an organised structure to deal with it. This is what motivated the authors to prepare a framework. In the framework, each block is based on some rational questions on the basis of which required action is to be taken to fulfil the resilient grid objective.

There are two ways of preparing a framework; one is strategic in which resources are managed for desire outcome and another is operational based on available resources outcome. In this chapter, mostly strategic perspective is considered because economies like United States, India and China not only have wide resources but also have the capability to manage the required resources.

A. **Events:** Background: What are the events (primary causes of outage like tropical storm) for which the utility has a prepared response? How these events were selected (experience, expertise, tradition, regulatory requirements, design basis, risk assessment, industry standard, etc.)

B. **Resource allocation**: In this section, the resource allocated for the event response is to be studied. Budgets, number of generators available, manpower, smart grid tech and questions related to monitoring system can be asked in this section. This section utilities special arrangement for a particular event is of major attention

C. **Capacity and capability**: Capacity defines the organised structure of the allocated resource. The type and intensity of the damage to the grid depend upon kinds of event. The capability of allocated resource to handle the event specific damage to grid is desirable. For the event of storm, the required capabilities for response is different than in the event of cyber attacks.

D. **Threshold**: When is a response activated? What is the triggering criterion or threshold? Is the criterion absolute or does it depends on internal/external factors? Is there a trade-off between, for example, safety and productivity? Are some important that is associated with this section.

E. **Performance and verification**: This block is dedicated to the analysis of the produce result. After an event recovery does not imply perfect restoration of the system functionality rather it implies that system has returned to a state where it is considered functional. Efficiency, reliability, fault tolerance and robustness are some performance characteristic measures, which are helpful for data verification.

F. **Outcome**: Whether the performance through prepared response is acceptable. How much system generated the outcome that was it seeking to achieve. Reduced damage from disasters and increased economic activity are some measures of outcome.

G. **Relevance**: This is the beauty of proposed framework which allow utilities to raise question upon its preparation. The most important question is that when the response was prepared, and when was it last updated? And under which circumstances?

A single event has the capability to create major disturbances in the grid but blackouts are not the results of single event and deficiency but the combination of several deficiencies. The following preconditions are the basis for a high power outage risk:

• High loadability limit or high grid utilisation
• For better unit commitment and optimal power flow
• Defects due to ageing infrastructure

Under these abovementioned conditions grid vulnerability increases. Now the likelihood of a power blackout for the following events is very high:

- Power plant shutdown due to capital maintenance or preventive maintenance or due to supply failures (e.g. cooling water shortage during heat waves)
- Operating personnel failure during maintenance work or switching operations
- Simultaneous grid interruption, for example, short circuit caused by tree contact, excavation work, balloons drifting into power lines, cars hitting utility poles, provisional shutdown due to electrical overloading risk
- Sudden demand of load growth
- Power line and its related electrical equipment may undergo outage
- Under breakdown due to natural calamities (e.g. storm, earthquake, snow or ice load, flood, lightning, extreme temperatures)
- Lack of communication between transmission/distribution system personnel and power generation company
- Intentional cyber attacks

7.3 RELATIONSHIP BETWEEN RESILIENCY AND PERCOLATION THRESHOLD

The ability of the power system to resist, mitigate and overcome stresses, failures and their consequences is receiving increasing attention from managers and system engineers with robustness, and resilience is gradually becoming very important considerations along with the more typical considerations of low cost and reliable operation. Moreover, proper understanding of the vulnerable and critical locations in transmission systems provide invaluable information that may be used to inform power grid management practices and programmes, leading to more realistic risk assessments and the development of defensive strategies to ensure network survival in case of extreme events and natural or man-made disasters.

Modern power grid is highly interconnected and plays a vital role in fuelling infrastructure and electric markets. Any disruptions in any components of power system, which results in cascading failures due to natural disasters, such as acts of war and terrorism, can ripple through the entire power system magnifying the original damage. Even relatively minor disturbances, such as congestion of transmission lines, can result in disproportionately severe disruptions to the electric power industry. Large centralised power generation sources, transmission lines, substations, especially those with high-voltage transformers, control centres are potential targets for terrorists and are vulnerable.

Network structure, network dynamics and failure mechanism play important roles in determining the resilience of realistic critical infrastructures, which may also shed light on the study of the resilience of the power grid. The characterisation of resilience is the main motivation for the studies involving complex network analysis and power grid [10]. The behaviour in terms of connectivity of the network when nodes or edges are removed is the primary question in many works considering failures that happen in a random fashion or following an attack. In general, the reliability is

assessed by evaluating the connectivity or the ability to efficiently guarantee paths
between nodes when nodes or edges in the network are removed [27]. Consequently,
for failures related to nodes, all the samples show a good resilience to random breakdowns. In fact, the network is always able to guarantee certain connectivity until the
numbers of nodes removed are the biggest part of it. On the other hand, the grids
are extremely vulnerable to targeted attacks, that is, failures that focus on key nodes
for the entire network such as high degree nodes or nodes with high betweenness or
nodes or lines that manage the highest amount of load or electricity flow.

Intentional attacks strongly deviate from random failures: even a small fraction
of removed nodes having large degrees has dramatic consequences. To predict the
effects of such directed attacks on network structure, the critical probability associated to network breakdown can be computed.

The probability of achieving connectivity from the source nodes to the terminal
nodes gives the reliability of the network. When terminal connectivity is concerned,
the identification of operational limits of a network goes missing where a critical
fraction of functional components to sustain the network is considered instead of
studying paths in the terminal reliability. This gap is overcome by percolation theory
which provides an opportunity by referring network failure to the situation whereby
a critical fraction of network components have failed.

In percolation theory, the failure of a node/edge of network is modelled by
removal. As the removal of nodes/edges increases, the network undergoes a transition from the phase of connectivity (functional network) to the phase of discon-
nectivity (nonfunctional network). The probability threshold signifying this phase
transition can be found theoretically or computed numerically by percolation theory.
The probability threshold can be used as a statistical indicator for the operational
limits of the network, which is not considered in traditional terminal reliability analysis. Thus, percolation theory, based on statistical physics, can help to understand the
macroscopic failure behaviour of networks in relation to the microscopic states of the
network components.

7.4   POWER SYSTEM BEHAVIOUR AT PERCOLATION THRESHOLD

In application of complex network theory for an electric grid network, the grid buses
are abstracted as nodes and transmission lines as links [28]. During operation at
steady-state condition, the nodes are fitted with a probability of $\rho = 1$, ρ being the
probability that a node is functional [27]; however, when the system encounters a
contingency or an attack that would affect the nodes of the power system, the probability of each node to remain functional would become a fraction <1 and would
have the probability $(1 - \rho)$ to be nonfunctional where $\rho < 1$. The *threshold value*
of probability of each node being functional under attack is termed as *percolation
threshold* ρ_c [27]. It can be used to obtain the critical fraction of nodes that can afford
to be damaged from any contingency or attack. At $\rho = \rho_c$ the connectivity reaches the
point of criticality following the damages sustained by the entire network. Whenever
$\rho > \rho_c$ there would be connectivity for the respective nodes.

Once the percolation threshold is obtained for the grid network under attack, it
is then possible to determine whether at least one path remains connected with the

main power network when the network is subjected under contingency or attack condition. Molloy–Reed criterion [29] being an effective tool to determine the percolation threshold of the nodes in a complex network, it can be applied in assessing the resiliency of the grid network under stress. Following this criterion, the percolation threshold transition is observable when $<c^2> = 2<c>$, where $<c^2>$ is the second moment of degree distribution and $<c>$ is the first moment of degree distribution. When $<c^2> \rightarrow 2<c>$, the power grid network transforms its nature from a nonresilient system to a resilient one in the limiting case.

Once the threshold of the grid network is established, it would be feasible to assess the resiliency of a specific structure of the grid network so that at least one critical load is catered (which may be used as a resource point to restore other loads affected by the vulnerability of any attack). In case the fraction (r) of the affected nodes of the grid network becomes lesser than the critical fraction r_{cri} of the quantum of affected nodes of the grid network, the grid system would have better resiliency. In simulating the power network, encountering attack or contingency, it would be useful to determine r_{cri} of the nodes that can be afforded to get damaged while maintaining the resiliency.

While encountering an attack or facing a contingency, the network graph configuration gets altered. This altered graph of the network is then termed as damaged graph. The damaged graph is characterised by the following degree distribution $P(c)$ [30,31]:

$$P(c) = \sum_{i \geq c}^{\infty} \binom{i}{c} r^{i-c} (1-r)^c P(c) \tag{7.1}$$

The standard generating function methodology employed to study the percolation properties identifies the two first generating functions of the damaged graph which can be represented as:

$$F_0(x) = \sum_{c}^{\infty} P(c)(1-r)x^c \tag{7.2}$$

$$F_1(x) = \frac{1}{c} \sum_{c}^{\infty} c\, P(c)(1-r)x^{c-1} \tag{7.3}$$

Here, $F_0(1)$ is the fraction of nodes from the original graph belonging to the damaged graph and $F_1(1)$ is the relation among $\langle c \rangle$ and the average number of nodes from V that can be reached after deleting a fraction r of nodes. The generating function of other components which can be reached from a randomly chosen node is:

$$H_1(x) = r + xF_1(H_1(x))s \tag{7.4}$$

And the generating function for the size of the component to which a randomly chosen node belongs to is [30]:

$$H_0(x) = r + xF_0(H_1(x)) \tag{7.5}$$

Thus, the average component size other than the giant component will be:

$$\langle s \rangle = H_0'(1) = 1 - r + F_0'(1) \times H_1' \qquad (7.6)$$

After long algebra, we see that this leads to singularity when $F_1'(1) = 1$

To ensure the percolation of the damaged graph, the following inequality holds good:

$$\sum_c (c-2) P(c) > \sum_c (c-1) \; r \; P(c)$$

$$\qquad (7.7)$$

$$\text{Or, } \langle c^2 \rangle - 2\langle c \rangle > r \left(\langle c^2 \rangle - \langle c \rangle \right)$$

Hence, it is possible to establish [32] equations given in (7.8) and (7.9).

$$\langle c^2 \rangle = \sum_c c^2 P(c)$$

$$\qquad (7.8)$$

$$= (1-r)^2 \langle c^2 \rangle + r \, (1-r) \, \langle c \rangle$$

$$\langle c \rangle = \sum_c c P(c) = (1-r)\langle c \rangle \qquad (7.9)$$

Application of Molloy–Reed criteria [30] in solving Equations (7.5) and (7.6) yields the expression for critical fraction of node removal r_{cri}, as shown in Equation (7.10),

$$r_{cri} = 1 - \cfrac{1}{\left(\cfrac{\langle c^2 \rangle}{\langle c \rangle} \right) - 1} \qquad (7.10)$$

Hence, it is evident from Equation (7.10) that the critical ratio of damaged to undamaged nodes in the given power network (r_{cri}) is governed by the ratio of variance (c^2) and average degree distribution (c) for the network configuration under consideration.

When the variance in the degree distribution gets maximised, the grid network becomes more resilient. It may also be observed that even if average degree connectivity increases without increase in variance, it would not improve the resiliency of the grid network. On the other hand, when both variance and average degree connectivity increases, the network resiliency will be more prominent. It is possible to compute the critical probability associated with grid contingency to predict the consequences of contingency or attacks on the grid network structure. It is possible to translate a contingency or an attack into equivalent random failure [31]. This gives

$$\rho' = \int_c^{c'} \frac{c \, P(c)}{\langle c \rangle} \qquad (7.11)$$

Here, ρ' represents the probability that a particular line leads to a deleted node (bus) following the removal of a fraction r of nodes (buses). This becomes equivalent to the random removal of those links (lines) interconnecting the remaining nodes to those already deleted.

The probability of having a node linked to other nodes can be represented as

$$P(c) = \exp\left(-c'\right)\big/\gamma \tag{7.12}$$

γ is a constant that characterises the exponential distribution corresponding to the average degree c of the undamaged graph. This results:

$$\rho' = \left(\frac{C'}{\gamma} + 1\right) e^{-\frac{c'}{\gamma}} \tag{7.13}$$

where C' represents the average degree for the damaged graph (network under attack).

7.5 ASSESSMENT OF RESILIENCY FOR TRANSMISSION NETWORK

The electric power transmission system, which is identified as a critical infrastructure across the world, possesses all the characteristics of a complex network. With ongoing smart grid projects implementation and research activities, power systems have to interact more actively with consumers, internet-based communication networks and transportation networks. This essentially transforms it into a "complex network of complex networks" with failure in one network affecting the entire system. Algorithms for fast optimal restoration of system loads are well known and widely used in industry. Though such restoration strategies are indispensable for continuity of power supply, self-organising nature of systems may conceal design flaws in the transmission systems which may determine over the long run; for example, certain transmission system reconfiguration may make the transmission system more vulnerable to cyber terrorism. Complex network analysis of transmission system resiliency is not to be interpreted as an attempt to improve upon existing (and successful) reconfiguration strategies; instead, it offers a bigger picture which may aid in understanding transmission and better operation and planning.

Power transmission systems are characterised by the presence of heavily connected branching nodes, where several buses feed off a long secondary and tertiary feeder lateral. A terror attack or weather-based damage, aimed at knocking down those branching nodes, has consequences for the continuity of power supply to consumers.

In this section, the focus is on the idea that a deletion of a very small fraction of nodes is enough to disintegrate the network down into islands. The damage extends from possible limitation of reconfiguration algorithms to include these "islands" [33] to cause a cascading outage leading to blackouts. A complex network analysis of these blackouts is well documented in [34].

Percolation theory has been found to be useful for studying the behaviour of physical systems characterised by random configurations [35]. In context of transmission systems, it is appropriate to consider the.flow of electrical energy through the network. A statistical physics backed percolation model served as an abstraction of complex networks and connectivity between the nodes [36]. Thus, this theory comes across as suitable candidate for analysis of resiliency of transmission system. Some of the definitions of system resiliency derived from the statistical tools employing complex network theory [37] are given below:

- A power network is resilient enough to maintain connectivity of one critical load to the main grid if **the probability of random damage of node being functional in the event of disruptive event is higher than the percolation threshold for the network.**
- A power network is resilient enough to maintain connectivity of one critical load to the main grid if the **second moment of degree distribution of nodes in the distribution system is greater than twice the first moment of degree distribution of the network configuration.**

7.6 SIMULATION

To investigate the resilience characteristics of a large transmission system against any attack or outage, the developed concept has been validated in the part of the eastern grid system of India (203 bus, 267 lines and 24 generator electrical grid network) [38]. The following algorithm illustrates the process of testing grid resilience against possible attacks and outages.

Step-by-step methodology:

1. The power network has been modelled as an undirected graph. A generation dispatch process is performed with generation-load balance using optimal power flow programme employing Newton–Raphson method in grid systems under consideration. The output exhibits parameters of grid operation including power flow in the lines.
2. The degree/connectivity, average degree distribution and variance (second moment of average degree distribution) of the network are determined.
3. The electrical betweenness of the nodes and lines of the network are determined.
4. With the highest betweenness generator buses being targeted for attacks, their resilience is analysed by determining the critical fraction (r_{cri}) of damaged nodes and probability (ρ_c) of the adjacent load nodes when the MW delivery of these generators are controlled from their full MW generation till outage (in five steps). Respective graphs of r_{cri} and ρ_c are plotted showing the pattern of these parameters with decrease in generation.
5. Next, the highest betweenness lines are analysed. The critical fraction (r_{cri}) of damaged nodes and probability (ρ_c) that a specific link leads to a deleted node are then calculated following removal of these lines. Again graphs of r_{cri} and ρ_c are plotted for the lines after their removal.

In this simulation, in the first step, the electrical betweenness of the nodes (buses) and links (lines) are determined. In the following section of the simulation, generator buses are ranked in descending order of electrical betweenness (Table 7.1). Here the generator full load ratings (in p.u.) are taken into account, and according to ranking of the electrical betweenness of these generator buses, the p.u. power ratings are curtailed from full load value till zero (complete outage). The corresponding magnitudes of r_{cri} and ρ_c are computed and the obtained values are tabulated in Table 7.1.

It may be observed from Figure 7.2 that following generation drop of the generators in the transmission system, the magnitude of r_{cri} and ρ_c reduce gradually indicating drop in resiliency of the grid. The grid becomes vulnerable to outage and attacks. Interesting to note how the values of r_{cri} and ρ_c reduce gradually, indicating the system is falling under poor resiliency. It may be observed that the slope of profile of r_{cri} and ρ_c corresponding to generators at Farakka are steepest for the major portion of operation, indicating that any attack or outage on generators at Farakka would make the transmission system least resilient. It is also evident that generators at Bakreshwar follows next. Hence, it is possible to identify specific generating station that would adversely affect the resiliency of the transmission network the most by observing r_{cri} and ρ_c profiles.

TABLE 7.1
Computational Values of r_{cri} and ρ_c Following Gradual Generation Drop of Highest Betweenness Generator Buses in Descending Order

Generator Bus Number	Generation (p.u.)	Highest Betweenness Load Bus Number Adjacent to the Generator	r_{cri}	ρ_c
1 (FARAKKA)	4	47 (BALURGHAT)	0.64	0.44
	3		0.49	0.38
	2		0.32	0.21
	1		0.28	0.12
4 (BAKRESHWAR)	1.9	188 (SAINTHIA)	0.83	0.72
	1.5		0.74	0.36
	1		0.68	0.29
	0.5		0.6	0.23
3 (KTPS)	1.8	71 (KOLAGHAT)	0.78	0.38
	1.4		0.73	0.32
	1		0.69	0.29
	0.5		0.66	0.23
8 (DPL)	0.9	199 (BOLPUR)	0.64	0.79
	0.75		0.60	0.43
	0.5		0.58	0.38
	0.25		0.55	0.22
13 (CHUKHA)	0.675	114 (COOCHBEHAR)	0.63	0.36
	0.5		0.6	0.29
	0.35		0.58	0.21
	0.2		0.55	0.18

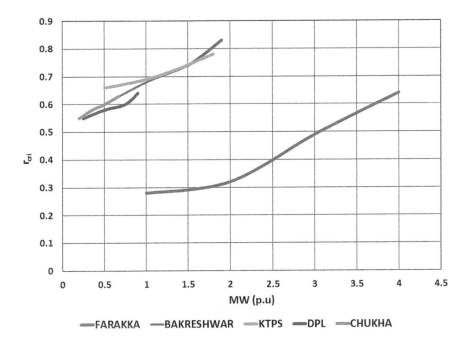

FIGURE 7.2   Variation of r_{cri} with change in p.u. generation for the respective generating plants.

Figures 7.3 and 7.4 represent the bar diagrams for r_{cri} and ρ_c of generators at respective generating stations arranged from the value of highest to lowest betweenness. It is evident from the simulation that generators at Farakka have the lowest values of r_{cri} and ρ_c indicating their susceptibility to attacks and outages followed by generators at Bakreshwar.

Next, the resiliency of lines has been studied by ranking the lines according to electrical betweenness in descending order and then removing those selected lines the values of r_{cri} and ρ_c are computed. It has been observed that line number 169, 241, 245, 205, 106, 148 and 42 are the first seven high betweenness lines. Outage of any or pair of these lines would make the drop in resiliency for the entire network. Table 7.2 exhibits the respective values of r_{cri} and ρ_c on removal of each of high betweenness lines in the given network.

The pattern of the values of r_{cri} and ρ_c when arranged according to the electrical betweenness of selected lines are shown in Figures 7.5 and 7.6, respectively. It is observed that as the electrical betweenness lines of the lines decrease, the value of r_{cri} and ρ_c increases. It signifies that for higher values of electrical betweenness the damage is much larger in proportion to those lines with lower electrical betweenness. It is evident that r_{cri} and ρ_c values of the lines indicate how much resilient is the network against tripping of any particular line.

The parameters r_{cri} and ρ_c depict the pattern of resiliency of a transmission network taking into account the possible outage of the generators as well as transmission lines

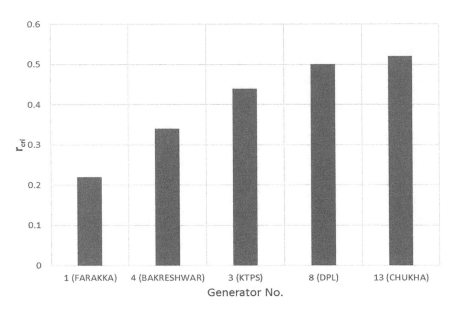

FIGURE 7.3   Bar diagram of r_{cri} for generators arranged from the highest value of betweenness to the lowest value of betweenness.

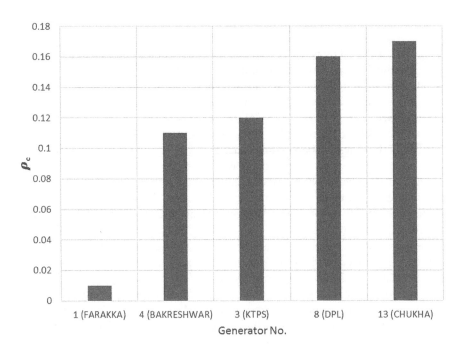

FIGURE 7.4   Bar diagram of ρ_c for generators arranged from the highest value of betweenness to the lowest value of betweenness.

TABLE 7.2

Results on Removal of Highest Betweenness Lines

Line No.	On Removal	
	r_{cri}	ρ_c
169 (FKK400-FKK220)	0.23	0.01
241 (FKK400-220-FKK33)	0.28	0.03
245 (117KTPS220-132-KTPS133)	0.32	0.07
205 (BKR15.75-BAKR400)	0.35	0.14
106 (HALD-TAML)	0.42	0.18
148 (HALD-HPCL)	0.47	0.21
42 (CHUKHA-BIRPARA)	0.49	0.23

with descending order of betweenness. This clearly signifies the relation of electrical betweenness in measuring the criticality of grid resilience. The part of the eastern grid system of India under study is basically a medium-to-large transmission system and the assessment of its resiliency gives an idea of how structural topological parameters like electrical betweenness are essential to maintain resiliency of transmission systems.

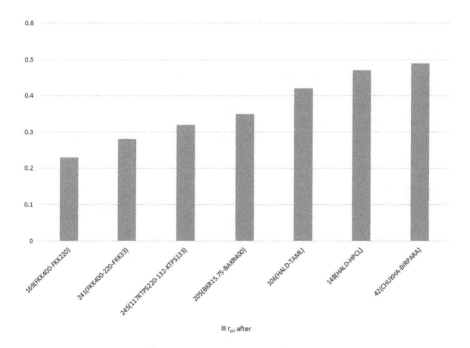

FIGURE 7.5 Bar chart showing the values of r_{cri} for selected lines for higher to lower betweenness.

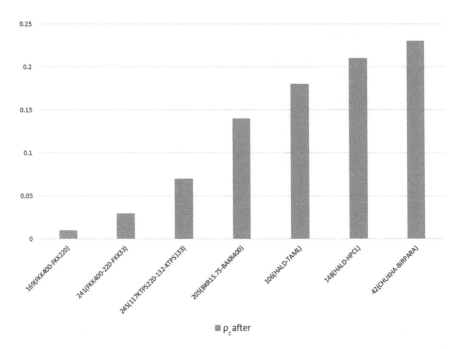

FIGURE 7.6 Bar chart showing the values of ρ_c for selected lines for higher to lower betweenness.

7.7 SUMMARY

The ability to resist and mitigate stresses, failures, outages or any attacks can be assessed by assessing the critical fraction of nodes and percolation threshold. These parameters are observed on removal of lines and buses; hence, resiliency is ascertained in transmission systems. It is possible to identify any specific generating station or transmission line whose removal would amount to the lowest resiliency of the transmission network.

REFERENCES

1. D.L. Alderson, G.G. Brown, and W.M. Carlyle, "Operational models of infrastructure resilience", *Risk Analysis*, vol. 35, pp. 562–586, 2015.
2. L. Xu, D. Marinova, and X. Guo, "Resilience thinking: A renewed system approach for sustainability science", *Sustainability Science*, vol. 10, no. 1, pp. 123–138, 2015.
3. D. Woods, "Four concepts for resilience and the implications for the future of resilience engineering", *Resilience Engineering and System Safety*, vol. 141, pp. 5–9, 2015.
4. D. Helbing, "Globally networked risks and how to respond", *Nature*, vol. 497, no. 7447, pp. 51–59, 2013.
5. L. Howell, *"Global Risks"*, Eighth Edition, Switzerland: World Economic Forum, 2013.
6. L. Mili, "Taxonomy of the characteristics of power system operating states", *In 2nd NSF-VT Resilient Ans Sustainable Critical Infrastructures (RESIN) Workshop*, Tucson, AZ, pp. 13–15, 2011.

7. P. Kundur, J. Paserba, and V. Ajjarapu, "Definition and classification of power system stability IEEE/CIGRE joint task force on stability terms and definitions", *IEEE Transactions on Power Systems*, vol. 19, no. 3, pp. 1387–1401, 2004.

8. R.J. Campbell, "Weather related power outages and electric system resiliency", *Congressional Research Service, Library of Congress*, USA, 2012.

9. C. Yijia, S. Xiaogang, and C. Ke, "Structural vulnerability analysis of large power grid based on complex network theory", *Transactions of China Electro Technical Society*, vol. 10, p. 26, 2007.

10. G.A. Pagani and M. Aiello, "The power grid as a complex network: A survey", *Physica A*, vol. 392, pp. 2688–2700, 2013.

11. United States Presidential Policy Directive-21, "Critical infrastructure security and resilience", vol. 12, 2013.

12. R.H. Lasseter, "Microgrids", *In IEEE Power Engineering Society Winter Meeting 2002*, vol. 1, pp. 305–308, New York, 2002.

13. US Department of Housing and Urban Development, "Hurricane Sandy rebuilding strategy", Washington DC, 2013.

14. R.C. Van Den Honert and J. McAneney, "The 2011 Brisbane floods: Causes, impacts and implications", *Water*, vol. 3, no. 4, pp. 1149–1173, 2011.

15. B. Zhou, et al., "The great 2008 Chinese ice storm: Its socioeconomic-ecological impact and sustainability lessons learned", *Bulletin of the American Meteorological Society*, vol. 92, no. 1, pp. 47–60, 2011.

16. C.S. Holling, "Resilience and stability of ecological systems", *Annual Review of Ecology and Systematic*, vol. 4, pp. 1–23, 1973.

17. C.S. Holling, L.H. Gunderson, and S. Light, *Barriers and Bridges to the Renewal of Ecosystems*, New York: Columbia University Press, 1995.

18. B. Walker, S. Carpenter, J. Anderies, N. Abel, G.S. Cumming, M. Janssen, L. Lebel, J. Norberg, G.D. Peterson, and R. Pritchard, "Resilience management in social-ecological systems: A working hypothesis for a participatory approach", *Conservation Ecology*, vol. 6, no. 1, p. 14, 2002.

19. J.M. Kendra and T. Wachtendorf, "Elements of resilience after the world trade center disaster: Reconstituting New York City's Emergency Operations Centre", *Disasters*, vol. 27, no. 1, pp. 37–53, 2003.

20. G. Brown, M. Carlyle, J. Salmerón, and K. Wood "Defending critical infrastructure," *Interfaces*, vol. 36, no. 6, pp. 530–544, 2006.

21. D.A. Reed, "Electric utility distribution analysis for extreme winds," *Journal of Wind Engineering and Industrial Aerodynamics*, vol. 96, no. 1, pp. 123–140, 2008.

22. G.P. Cimellaro, A.M. Reinhorn, and M. Bruneau, "Framework for analytical quantification of disaster resilience," *Engineering Structures*, vol. 32, no. 11, pp. 3639–3649, 2010.

23. T. Aven, "On some recent definitions and analysis frameworks for risk, vulnerability, and resilience," *Risk Analysis*, vol. 31, no. 4, pp. 515–522, 2011.

24. D. Henry and J.E. Ramirez-Marquez, "Generic metrics and quantitative approaches for system resilience as a function of time," *Reliability Engineering and System Safety*, vol. 99, pp. 114–122, 2012.

25. E. Hollnagel, D.D. Woods, and N. Leveson, *"Resilience Engineering: Concepts and Percepts*, Farnham: Ashgate Publishing Ltd., 2007.

26. E. Zio, "Reliability engineering: Old problems and new challenges", *Reliability Engineering and System Safety*, vol. 94, no. 2, pp. 125–141, 2009.

27. D. Li, Q. Zhang, E. Zio, S. Havlin, and R. Kang, "Network reliability analysis based on percolation theory", *Reliability Engineering and System Safety*, vol. 142, pp. 556–562, 2015.

28. H. Ahmadi and J.R. Marti. "Mathematical representation of radiality constraint in distribution system reconfiguration problem", *International Journal of Electrical Power and Energy Systems*, vol. 64, pp. 293–299, 2015.

29. M. Molloy and B. Reed, "A critical point for random graphs with a given degree sequence", *Random Structures and Algorithms*, vol. 6, no. 2–3, pp. 161–180, 1995.

30. R. Cohen, K. Erez, D. Ben-Avraham, and S. Havlin, "Resilience of the Internet to random breakdowns", *Physical Review Letters*, vol. 85, no. 221, p. 4626, 2000.

31. R.V. Sole, M. Rosas-Calas, B. Corominas-Murtra, and S. Valverde, "Robustness of the European power grids under intentional attack", *Physical Review E*, vol. 77, pp. 026102–26109, 2008.

32. J.H. Grisi-Filho, R. Ossada, F. Ferreira, and M. Amaku, "Scale-free networks with the same degree distribution: Different structural properties", *Physics Research International*, vol. 2013, pp. 2090–2220, 2013.

33. C. Wang, S. Ruj, M. Stojmenovic, and A. Nayak, "Modelling cascading failures in smart grid using interdependent complex networks and percolation theory", *IEEE 8th Conference on Industrial Electronics and Applications (ICIEA)*, pp. 1023–1028, Melbourne, 2013.

34. P. Crucitti, V. Latora, and M. Marchiori, "Model for cascading failures in complex networks", *Physical Review E*, vol. 69, p. 045104, 2004.

35. R. Pastor- Satorras and A. Vespignani, *"Evolution and Structure of the Internet: A Statistical Physics Approach"*, Cambridge: Cambridge University Press, 2007.

36. Y. Zhang and L. Guo, "Network percolation based on complex network", *Journal of Networks*, vol. 8, no. 8, pp. 1874–1881, 2003.

37. S. Chanda and A.K. Srivastava, "Quantifying resiliency of smart power distribution systems with distributed energy resources", *2015 IEEE 24th International Symposium on Industrial Electronics (ISIE)*, Rio de Janeiro, Brazil, 2015.

38. K. Chakrabarti, "Investigation on voltage security of power transmission systems using soft computing techniques", PhD Thesis, BESU Shibpur, 2012.

8 Effect of Distributed Energy Sources

8.1 INTRODUCTION

Distributed renewable energy sources (like wind turbine, photovoltaic, fuel cell, biomass, smart house, etc.) and energy storage devices (like battery, electric double layer capacitor, superconducting magnetic energy storage, etc.) are the future of energy demand [1]. Distributed generation (DG) located close to the load, that is, on the distribution network or on the customer side of the metre, have great potential to improve distribution system performance and should be encouraged.

DG [2] refers to small electric power generators, typically ranging in capacity from 1W to 300 MW, which can be located on the utility system, at the customer site or at a location not connected to the grid. DG can be conventional, such as combined cycle turbines, small diesel generators, combustion turbines or wind turbines, solar generation and other renewable energies, which will be discussed in detail in this chapter.

Rating of DGs*:* The maximum rating of the DG which can be connected to a distributed generation depends on the capacity of the distribution system that is interrelated with the voltage level of the distribution system. Hence, the capacity of DGs can vary widely. There are four different categories of DGs [1]:

Micro: DG range: ~1W<5kW;
Small: DG range: 5 kW<5MW;
Medium: DG range: 5 MW<50MW;
Large: DG range: 50 MW<~300MW.

Because of the various types of DGs, the generation electric current can be either direct current (DC) or alternating current (AC). Photovoltaic, fuel cell and batteries generate DC which is appropriate for DC loads and DC SG (Smart Grid). On the other hand, the DC can be converted to the AC using power electronics interface, which can be connected to the AC loads and power grid. Other DGs such as wind turbine, micro turbine and biomass deliver an AC, which for some applications must be controlled using modern power electronic equipment to acquire the regulated voltage [1].

8.1.1 SOLAR PHOTOVOLTAIC ENERGY

The basis of solar photovoltaic energy conversion lies in the concept of how light energy is directly converted to electrical energy. Sunlight consists of packets of energy, also known as photons. This photon inherently possesses energy, and when it falls on PV cell it can liberate electrons. The energy of photons liberates electrons from the valence band to the conduction band. and creates electron hole pair. One photon is responsible for one electron-hole pair. The energy of photon depends on the wavelength of the light.

$$E = \frac{hc}{\lambda}$$

where h = Planck's constant, c = speed of light and λ = wavelength.

The energy of photon must be equal to the energy gap of that material.

The effectiveness of PV conversion depends upon many factors:

- Surface properties like absorptivity and reflectivity
- Rate of charge generation and charge recombination within semiconductor material

In general, silicon (energy gap is 1.107 eV) is used as the material of PV cell but gallium arsenide (GdAs), indium phosphide (InP), cadmium telluride (CdTe), etc. can also be used.

A solar cell has limited open circuit voltage and current delivering capability. To increase its power rating, they are connected in suitable series-parallel combination in an enclosed chamber, known as a solar module or solar panel. In a similar manner, solar modules are connected on load-specific application-based series-parallel combination to form a solar array. A solar panel is rated by its DC output power at a certain voltage at SOCT(standard operating cell temperature). At a higher temperature, the efficiency of solar cell decreases; hence, it is recommended to provide better ventilation in any solar module. During the formation of a solar array, a designer should care about mismatch loss due to interconnection of panels having different properties. It is an undesired phenomenon as it can create a localised circulating power flow reducing the desired output of a solar cell to a considerable extent.

Another problem regarding PV operation is shading. When one solar cell is covered or shaded by any means such as a leaf or a shadow, the covered portion heats up to such an extent that the cell or module concerned may get damaged. This defect is called a hot spot. A shaded cell does not produce electricity but consumes current from the sound ones. Hence, the current from the string is converted into heat. To protect a module from hot spot, bypass diodes are used. Their function is to provide an alternating path to the current concerned in shaded cell condition. To avoid most of the unwanted situation and to maximise solar insolation, solar PV-based systems are always installed in a place where the chances of shadow are a minimum. Even in array, the distance between the rows of solar modules are chosen such that the shadow of one module do not fall to any module on the next row. In Northern Hemisphere, it is recommended that any solar installation should face south at an elevation angle equal to the latitude of the concerned place.

For a reliable power supply through solar PV method, the following components are used. However, depending on specific application other components like inverter might be used.

i. PV module

PV module consists of solar cells arranged in a suitable fashion to suit for the voltage or current demand. Several modules are connected in appropriate series-parallel combination to increase the power range.

ii. Storage element (Battery):

In general, an off-grid system requires a storage system for utilising sun's energy in absence of sun light during the night. A battery is a good storage choice in such a case. The capacity of the battery depends upon the estimated load. The problem regarding battery storage system is that the life period of a battery is much less than a solar module, so proper designing involves choosing the proper capacity of battery from an economical viewpoint. During operation, it has to be kept in mind that the battery should not reach its threshold discharge level, otherwise its lifetime will be shortened. For high-rating applications, several batteries are used in suitable series-parallel connection known as battery bank.

When a battery is a part of balance of system for PV-based supply is delivering current, it is not allowed to discharge to its full capacity. If this should be the case (i.e. battery is allowed to discharge fully), then its lifetime will be much shortened, and due to the high cost of battery, it becomes an uneconomical approach. To counter this problem, battery is allowed to discharge to a certain level (say 40%) depending on the type and manufacturer of the battery. This is known as Depth of Discharge (DOD). To maintain the DOD to the proper value, a charge monitoring unit, known as charge controller is used.

iii. Charge controller:

Basically, charge controller is a protective measure taken to protect the battery. The functioning of a charge controller is to protect the battery from undercharge or overcharge protection. In either of the two cases, this protecting device cut outs the load relieving the battery. Charge controllers might not be used in small-scale applications.

iv. Inverter:

The output of a solar photovoltaic system is always DC; however, most of the electrical systems encountered today is AC in nature. To encounter this problem an inverter is used at the end terminals of a PV system. In absence of an inverter such as for lighting or to run a fan separate line is laid out with different types of light or DC fan. Moreover, for a grid connected system, an inverter is essential.

Responsibility of a designer is not only limited for proper load matching and estimating the capacity or rating of different components. The designer should equally focus on the economics of the designed system. The target is to achieve a cost-benefit system. To avoid any tussle between these two contradictory requirements, the designer should choose an optimal design approach.

The cost of a PV system generally involves cost of it components such as PV array and its size, battery bank size and other component costs, which are application-specific.

Solar PV system can be generalised as generator side which consists of mainly PV array and accessories and balance of system, which includes load and other converting equipment (if any).Generator side consists of solar PV cell, modules or arrays and the remaining things are popularly known as balance of system. The power of module is generally given in Watt-peak or W_p, which is the amount of power generated by solar PV cell at 1,000 W/m^2 intensity at 25°C temperature at a air mass level of 1.5.

The designing process can be analysed as follows:

a. **Estimation energy demand and numbers of module required**: The first step involves calculating the energy required for a particular load profile. This involves calculating the total energy required for any particular day. Second, the total number of modules required is calculated based on the power rating of the module concerned. When estimating the size of a module, changes from the standard test condition to practical condition is done.

b. **Inverter sizing (if required)**: The output of a solar PV-based system is DC in nature. But nowadays generally all types of load are alternating in nature. To counter this problem, an inverter may be installed depending on the situation. Generally, the inverter size should be 25%–30% higher than the total power demand. In case any highly inductive machine operation is required in the load profile (such as compressor or motor), the inverter size should be minimum thrice the capacity of the total power demand to tackle the higher starting current. The input rating of the inverter should be the same as the PV array system for safer operation.

c. **Battery sizing**: Generally, the battery used for solar PV application is deep-cycle type. The reason for using this special type of battery is that they can discharge to a very low level. The capacity of the battery should be sufficient to supply load during the absence of sunlight. Sufficient measures are taken to consider battery losses (if any) and DOD. The battery voltage should match with the system voltage. Hence, required number of batteries are connected in series to provide the given voltage level. Similarly, to meet any specific demand, batteries are connected in parallel. This series-parallel combination is known as a battery bank.

Apart from the above items, few other items may be required on specific applications. For example, a charge controller may be required in few applications. The size of the charge controller depends on the current to be handled. But to provide better protection, that is, protection from short-circuit, the capacity of charge controller is chosen according to a factor (1.3–1.5) multiplied with the specified short-circuit current capability of the concerned system.

8.1.2 WIND ENERGY

Wind is the horizontal movement of air which arises due to uneven heating of the earth and derives its energy from solar radiation. At daytime, air above land masses heats up faster than air over the sea, oceans or any water bodies. Hot air expands and rises while cool air cools more rapidly over landmasses than water off shore land, causing land and sea breezes. Due to the shape of the earth, Equator receives the maximum heat causing winds to blow from subtropical belts towards the Equator. In addition, the axial rotation of the earth induces a centrifugal force which has a thrust on equatorial air masses to the upper atmosphere causing deflection of the winds.

The concept of harnessing wind energy dates back to Egyptian civilisation where they used wind power to sail their boats in the Nile River [3].Skilful technicians from the Middle East introduced windmills to China. The technology to harness wind energy reached Western Europe by the Arabs. Years have passed since then and wind power has emerged as one of the most used renewable energy across the world. After sudden price rise in fossil fuels, a number of countries were stimulated towards generation of power using wind energy.

Wind turbines extract energy from wind stream by converting kinetic energy of the wind to rotational motion required to operate an electric generator. By virtue of kinetic energy, the velocity of the flowing wind decreases. It is assumed that air masses which passes through rotor is only affected and remains separate from the air which does not pass through the rotor. As the free wind (stream) interacts with the turbine rotor, the wind transfers part of its energy into the rotor and the speed of the wind decreases to a minimum leaving a trail of disturbed wind (also called *wake*). The variation in velocity is considered to be smooth from far upstream to far downstream. The wind leaving the rotor is below the atmospheric pressure (in wake region) but at far downstream it regains its value to reach the atmospheric level. The rise in static pressure is at the cost of kinetic energy, consequently, further decreasing wind speed. Because wind flow is considered incompressible, air stream flow diverges as it passes through the turbine. Also, the mass flow rate of wind assumed constant at far upstream, at the rotor and at far downstream.

Basically, a wind turbine operates by slowing down the wind and extracting a part of its energy in the process. In general, wind turbines have blades, sails or buckets fixed to a central shaft. The extracted energy causes the shaft to rotate. This rotating shaft is used to drive a pump, to grind sees or to generate electric power. Power extraction by wind turbine depends on variation of wind turbine power with rotor diameter and wind speed. *Mean wind speed*, *energy estimation* and *power density duration curve* also play equally important roles.

Wind turbines are classified as horizontal-axis turbines or vertical-axis turbines depending upon the orientation of the axis of rotation of their rotors. Wind turbines are further classified into "Lift" and "Drag" type.

Although wind energy is the first among renewable energy to become an economically viable source, it is highly unpredictable depending on season, elevation of the land and wind characteristics like frequency of distribution of wind speed. Hence, assessment of wind energy potential has to be evaluated by recording meteorological parameters such as wind direction, wind speed temperature and rainfall on hourly basis. This data is collected from surveys and historical data are used for forecasting. For correct forecasting, the wind energy at a particular site is decided by the wind profile or regime (*wind speed frequency distribution*) of the site that can be mathematically described by *Weibull Probability Distribution Function*.

Many wind energy generators (WEGs) are commercially available in the capacity of 1 kW to 3 MW. It is necessary to select the best-suited WEG for a particular site for generating maximum energy. Production of electricity depends on mean wind speed, hub height, cut-in speed, rated and furling wind speed of the machine. Grid interfacing of a wind farm is also done. It can be integrated with state grid to ensure a smooth supply of continuous power. There are limitations on the integration of wind turbines to the grid system. Pumping the generated electricity from WEGs to state grid should have the minimum power quality impact on the grid so that integration should contribute to the improvement in power quality. *Grid short-circuit power* and *grid short-circuit ratio* have a great influence on the power quality of the grid.

A wind farm has several identical wind turbine generators. These are induction-type generators which need reactive power for magnetising. With conventional energy system, generators besides supplying active power, supply reactive power required by consumers to operate their electrical equipment. However, WEGs (induction type) require reactive power to start power generation. To have availability of reactive power, each WEG is provided with shunt capacitors. These capacitors meet reactive power requirement of WEG and maintain power factor at the rated value of 0.95. Induction generators can be used in both stand-alone and grid-connected mode with advantages such as low cost and robust construction. When connected with grid systems, WEG draws reactive volt-ampere from the grid. This VAR drains on the grid system which is compensated by the use of terminal capacitors.

8.1.3 SMALL HYDRO POWER

Falling water as a source of energy is known from old days where it was used to turn water wheels for grinding corn. Electric power is generated when water from a high altitude is made to flow through hydraulic turbines. Hydropower projects essentially harness energy from flowing or falling water in rivers, rivulets, artificially created storage dams or canals. Potential energy in water is converted into shaft work utilising a hydraulic prime mover. The hydraulic turbine converts potential energy of water or kinetic energy of flowing stream into mechanical energy by its rotating shaft. Electrical energy is obtained from an electric generator coupled to the shaft of the prime mover. It depends mainly on the head of the water and the discharge through the turbine. Hydropower projects of ratings <10 MW are regarded as small hydropower resources which are extremely favourable.

8.1.4 BIOMASS AND BIOGAS ENERGY

Biomass is organic or carbon-based matter (resources from forest, agriculture, aqua culture and residue from industry and urban waste) that can either react with oxygen for combustion or undergo metabolic process to release heat. Biomass can be used in original form or can be transformed to convenient and useful form of solid, liquid or gaseous. Biofuels can be fuel wood, charcoal, fuel pellets, bio-ethanol, biogas, producer gas and biodiesel. Biomass conversion technologies include *densification*, *combustion, incineration, thermochemical* (*pyrolysis, gasification and liquefaction*) and *biochemical conversion* (*anaerobic digestion* and *ethanol fermentation*). Biomass gasification can be done using *gasifiers* like *fluidised bed gasifier*.

Biogas is a gaseous flammable fuel (mainly composed of methane, carbon dioxide, hydrogen, hydrogen sulphide and other gases) which is obtained from biomass of an agricultural digestion or fermentation of wet organic matter. Biogas is produced through biological conversion process that involves anaerobic digestion proceeding subsequently as *hydrolysis, acid formation* and *methane formation.*

Rapid urbanisation and industrialisation has resulted in enormous quantity of waste in urban and industrial areas. *Municipal solid waste, municipal liquid waste* and *urban industrial wastes* that are disposed are also taken for energy conversion. Biomass and biogases are mainly used to power cooking and heating purposes primarily in rural areas.

8.1.5 TIDAL ENERGY

Tidal energy is another method of harnessing hydropower. Tidal power possesses a scope from power generation viewpoint in the coming future. Tide forecasting is more reliable than that for wind or solar. High initial cost and limited availability of sites having satisfactory tidal range has limited the scope of tidal energy extraction technique. At present, continuous research on plant and turbine design may prove to be economical.

Tides are formed by moon's gravitational pool on the oceans and at the estuaries of large rivers on the earth. The sun's pool in similar fashion to the moon plays a minor role in this case. Because of the strong attraction of the celestial bodies to the oceans, a bulge in the water level is created, causing a temporary increase in sea level. Consequently, sea level rises causing water from the middle of the ocean to be moved towards the shore areas, thereby creating a tide. This takes place in an unfailing manner because of the consistent pattern of the changing positions of the moon and sun relative to the earth.

During the rise and fall of sea level, the tidal generation system can be activated with proper gearing mechanism where necessary. The potential energy created by the pressure difference caused due to the water level difference across water passages and barrages is converted to kinetic energy in the turbines. This turbine is coupled mechanically with proper generator to produce electricity.

A tidal power plant consists of the following components:

a. Dam
b. Sluice gates from the basins to the sea
c. Power house

A dam is an artificial barrier between the seawater and the site of interest or basin. The primary function of a dam in tidal power plant is to absorb shock and pressure of wave. The dam provides proper channelling for turbine operation. Tidal barrages are constructed at suitable places where the head of water are satisfactory which is ~3 m. Choice of proper site for constructing a dam is a very difficult task because the energy available is related to the size of the basin (the area from dam to plant, precisely to turbine) and to the square of tidal range.

From the above discussion, it is clear that a dam should be built near to the estuary or bay. But the problem will be that with larger size of basin, the effective tidal range gets reduced. So, a balance must be maintained between these two contradictory criteria. Gates are used in tidal plant to fill and empty the basin as and when required. Due to the interaction with seawater, gates are prone to corrosion. This can be tackled by cathodic protection while constructing the gate. Power house is housed with the turbines, generators and other auxiliary instruments. Because of the small head, turbines of large sizes are required in this type of plants. Alternator is directly coupled to turbines. Generally, turbines used in hydro turbines with low head operation may be used in such a plant with proper modifications wherever necessary.

8.1.6 OCEAN THERMAL ENERGY CONVERSION (OTEC)

OTEC is a process that can produce electricity using the temperature difference between deep cold ocean water and warm tropical surface waters. OTEC plants pump large quantities of deep cold seawater and surface seawater to run a power cycle and produce electricity. OTEC is firm power (24/7), a clean energy source, environmentally sustainable and capable of providing massive levels of energy.

Recently, higher electricity costs, increased concerns for global warming and a political commitment to energy security have made initial OTEC commercialisation economically attractive in tropical island communities, where a high percentage of electricity production is oil based. Even within the United States, this island market is very large; globally, it is many times larger.

8.1.7 GEOTHERMAL ENERGY

Geothermal energy is the energy present in the interior of the earth. Geothermal energy can be extracted from earth's interior in the form of heat. Volcanoes, geysers and hot springs are visible signs of large amount of heat lying in earth's interior. The geothermal energy from earth's interior is almost inexhaustible; although the amount of thermal energy within the earth is very large, useful geothermal energy can be extracted at only certain suitable sites [4].

Geothermal resources are of five types: *hydrothermal (hot water and wet steam), vapour-dominated resource, hot dry rock resource, geo-pressured resource and magma resource.* The hydrothermal type and hot dry rock type are the most used. Hydrothermal plants generally have common features consisting of (i) production well to extract steam from the resource,(ii) a centrifugal separator to remove solid matter from the steam,(iii) a turbine to convert thermal energy to mechanical energy,(iv) a generator coupled to the turbine to generate electric power,(v) a condenser to condense wet steam exited from the turbine into water by direct contact with cooling water and (vi) a cooling tower to cool warm water exited from the condenser and returning the cooled water to the condenser.

In case of hot dry rock type, injection for wells pumping inside and production or extraction wells for hot water pumping out are drilled. A series of injection and extraction wells can be drilled to tap a sufficient amount of geothermal energy. The hot water extracted from the man-made reservoir is made to vaporise low boiling point refrigerant which is used to run a turbine coupled with a generator. The refrigerant vapour exiting the turbine is condensed in a condenser which is pumped into the heat exchanger again. The viability S of extracting energy from a dry field depends upon the degree to which the resource field can be fractured to develop man-made geothermal reservoirs.

8.1.8 MHD Generation

The magneto hydrodynamic (MHD) power generation technique is a direct method of energy conversion which converts heat energy directly into electrical energy, without any intermediate mechanical energy conversion, as is prominent in other power plants. Therefore, in this process, substantial fuel economy can be achieved due to the elimination of producing mechanical energy and then converting it to electrical energy.

The concept of MHD power generation was introduced for the very first time by Michael Faraday in the year 1832 in his Bakerian lecture to the Royal Society. He conducted an experiment at the Waterloo Bridge in Great Britain for measuring the current from the flow of the river Thames in earth's magnetic field. This experiment outlined the basic concept of MHD generation over the years then, and significant research work has been conducted on this topic. On August 13, 1940 this concept of MHD power generation was imbibed as the most widely accepted process for the conversion of heat energy directly into electrical energy without a mechanical sub-link.

The principal of MHD power generation is very simple and is based on Faraday's law of electromagnetic induction, which states that when a conductor and magnetic field moves relative to each other, voltage is induced in the conductor. As the name implies, MHD generator is concerned with the flow of a conducting fluid in the presence of magnetic and electric fields. In conventional generator or alternator, the conductor consists of copper windings or strips, while in an MHD generator, the hot ionised gas or conducting fluid replaces the solid conductor. A pressurised, electrically conducting fluid flows through a transverse magnetic field in a channel or duct. A pair of electrodes are located on the channel walls at right angle to the magnetic field and connected through an external circuit to deliver power to a load connected to it. Electrodes in the MHD generator perform the same function as brushes in a conventional DC generator. The MHD generator develops DC power and the conversion to AC is done using an inverter.

8.1.9 Energy Storage

At times the nature of the load is so fluctuating that there might be some instances when there is a mismatch between the generation and requirement. This problem can be met with the help of energy storage [5]. In other words, storage keeps energy in any suitable form within itself until it is required in the same or any other form. The storage element stores energy when generated amount is excess than the requirement and delivers the same when the demand is more than the generated amount. In case of electricity, it can be stated that the output in excess of load demand from any solar PV plant or any windmill is fed to any grid system. Otherwise, a storage element is required.

Basically, storage plays an important role in economic power generation. The main aim for combined operation of power plant is to minimise the operating cost with maximised output. This can be effectively done with the help of a storage system. The plants are operated optimally. The excess energy generated from this optimal schedule is stored in storage system which may be utilised during demanded period.

An important point to note on energy storage is that the nature of energy storage element is different in case of renewable energy resources than fossil fuel-based resources. This is because extraction rate of energy from most of the renewable energy-based resources (as an example we can think about a solar PV plant) is beyond human control unlike fossil fuel-based energy extraction techniques. This difficulty can be overcome by matching the load to the energy supply which is a very difficult task. Another option is to store energy for use in case of emergency.

8.2   ROLE OF DISTRIBUTED ENERGY SOURCES IN POWER TRANSMISSION NETWORKS

Although frequent occurrences of blackouts have been common since past few decades, in most of the studies, DG is not considered for vulnerability analysis of power grids. Actually, in many existing power grids, some load nodes are far away from the generators, so it is very difficult for these nodes to acquire enough power supply. In this case, the adoption of distribution generation technology will be a good choice. DG can reduce long-distance transmission and balance the power flow, making the power grid more reliable. In addition, DG can reduce the energy consumption and cost, enhancing both reliability and flexibility of power grids. Of course, failure of some part of the power grid with DG can also make the power grid become vulnerable, so the vulnerability assessment of power grid with DG is also equally related to this concern.

8.2.1   DGs in Restructured Environment

The DG placement problem can be formulated as an optimisation, with similarities to the optimal power flow problems. However, in practice, a network operator may only have a limited influence on the actual location of DG because this often depends on factors such as site availability, as well as construction and planning permission issues.

A lot of work has been done in the area of DG placement in the distribution network [6–8]. Among the few studies on the placement of DGs in the transmission system, [9] discusses the impact on grid dynamics if the grid is powered using DGs. They also assert that different dynamics can be seen on the grid by varying the fractions and distributions of DGs in the transmission system. They also mention that improper distribution and sizing of the DGs may lead to increased vulnerability of the grid instead of increased robustness. The stability and reliability of the grid improves with the use of conventional DGs close to the loads, but if an increasing penetration of stochastic renewable energy sources is present in the grid, these energy sources introduce erratic power inputs into the grid, thereby causing it to fail with a sharp transition [10].

The interconnection of several DGs in the transmission system causes a structural change in the grid, and, therefore, the properties of the system would change. Indices, such as characteristic path length, degree and degree distribution, clustering coefficient, and betweenness, decide how the interconnection of DGs aspects topological characteristics of the grid. They have also used weighted graph indices and have suggested new indices based on structure and operational conditions of the grid for the evaluation of structural properties of the grid with incorporated DGs.

8.2.2 Applications of DGs

There are several applications of DG in the power system [10]:

- The DG can be scattered in different places, and can be utilised as a standby power source. If the grid power cuts off the sensitive loads, for example, process industries and hospitals, the DG can provide the emergency power for these loads.
- The DG can supply power for geographically isolated communities which are difficult to connect to the main power grid. Therefore, the DG can improve the economic condition for isolated communities.
- The electric power cost depends on the electric load. When the load demand is high, the electric power price will be high and vice versa. The DG can supply the electric power to the load when the demand is high. Consequently, the customer can reduce the electricity cost to pay time-of-use rates.
- The DG can supply power for the rural and remote applications which include lighting, heating, cooling, communication and small industrial processes.
- Individual DG owners are usually used as a base load to provide part of the main required power and support the grid by enhancing the system voltage profile. The DG also helps to reduce the power losses and improve the system power quality.

DG affects the operation of the distribution networks, including power flows and voltages [1]. DG also has an effect on system losses. In many cases, the effect of DG on losses is positive as DG is often located close to the demands where the energy is consumed. This reduces the distance over which energy needs to be transported, therefore reducing losses.

DG can also impact the reliability and quality of the power supply in the electrical system. Previous studies have demonstrated some positive effects of DG on reliability, where DG can reduce the loading on the network at critical times, or supply part of the demand in the network during faults and shortages. However, DG can also have a negative impact on reliability, particularly in cases where it has caused problems for the coordination of the network protection systems [10].

8.3 MITIGATION STRATEGY EMPLOYING DISTRIBUTED ENERGY SOURCES

The use of DG in the sub-transmission and distribution system can be used to mitigate cascading failures causing large-scale blackouts. Despite the fact that DGs have their own arguments, their emphasis increases continuously because of their advantages like providing local power, thus inhibiting power transfer for long distances, which, on the other hand, improves grid reliability. This chapter focuses on allocation of DGs in the power networks. Renewable energy sources can be treated as a separate area of study. The placement of DGs is effected on the basis of relative criticality of nodes in the system employing the tool of *electrical centrality* which gives electrical node significance. The allocation of DGs strengthens the robustness of the grid network and obstructs cascading failures. Vulnerability analysis using power flow model is done prior to the allocation and post allocation of the DGs. The interaction of certain amount of DGs in a power network causes a topological change in the network, and, therefore, to what extent the properties of the network would change is also determined by assessing Grid Vulnerability Index (GVI).

8.3.1 NETWORK CRITICALITY ASSESSMENT USING [Z_{BUS}] CENTRALITY EMPLOYING DGS

The sub-transmission system has been receiving attention from complex network theory researchers for some time. The metric of electrical centrality which can be useful for installation of DG in distribution system and discussed in Chapter 4 differentiate the electrical topology and physical topology of power grid. This measure not only enables complex network analysis of power systems but is also more appropriate for the power grid than general topological analysis. This measure is based on Z_{bus} impedance matrix of a power system which finds more electrically central nodes in the system which is supposed to be highly connected to most other nodes for the placement of distributed generators in the distribution system.

8.3.2 INCORPORATION OF DISTRIBUTED ENERGY SOURCES IN CRITICALITY ASSESSMENT USING BETWEENNESS METRIC: SIMULATION FOR DG INCORPORATION IN CRITICALITY ASSESSMENT OF IEEE 57 BUS SYSTEM USING BETWEENNESS METRIC

The IEEE 57 bus system [11] has been considered to apply the concept presented here to study electrical betweenness of buses and check the ranking of their criticality. Figure 8.1 represents the topological network of IEEE 57 bus system.

Figure 8.2 represents the graphical plot of criticality magnitudes as obtained from the measure of betweenness of load buses. The electrical betweenness of load bus 11 being the maximum; it is followed by the load bus numbers 13, 16, 10. Thus, the most critical load bus is bus number 11 and it is vulnerable against any unplanned outage and attack. Following incorporation of DGs in the most critical load bus (i.e. bus 11), the criticality magnitude for this bus reduces. Thus, application of DGs reduces the

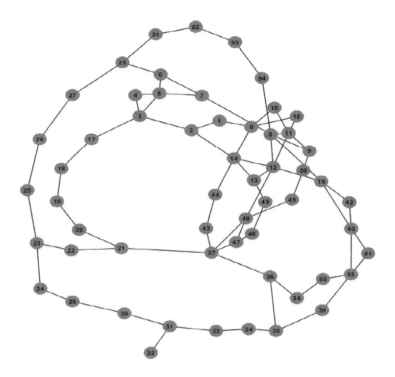

FIGURE 8.1   Complex network topology of IEEE 57 bus.

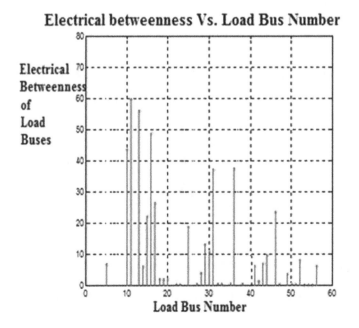

FIGURE 8.2   Electrical betweenness of load buses in IEEE 57 bus system.

criticality and ensures lesser vulnerability of most critical bus following unplanned outage or attack on the power network.

Figure 8.3 exhibits the decrement in magnitude of electrical betweenness of four successive load buses (which have high level of betweenness) following incorporation of DG at most critical load bus (i.e. bus 11 which has the highest magnitude of betweenness).

Load flow study was carried out after incorporation of DG to evaluate the losses. DG incorporation resulted in improved efficiency along with reduced losses and raised the voltage profile. It has been observed that the steady state voltage profile (Figure 8.4) of load buses improve with incorporation of DG while the power loss in the network reduce from 28.12 to 24.32 MW.

Hence, employment of DG unit not only reduces betweenness (criticality) but at the same time decreases system vulnerability, improves power transmission and increases voltage profile making the grid more efficient. The role of DG penetration in improvement of complex network theory parameters is an important tool which can be employed to strengthen the infrastructure of power grid.

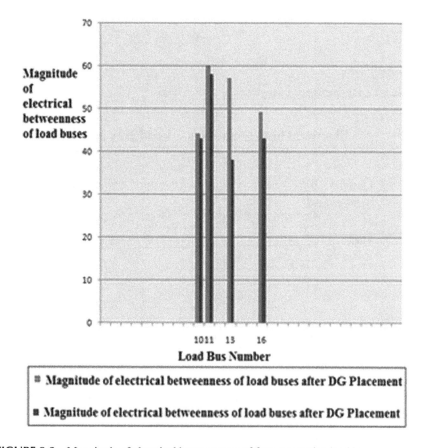

FIGURE 8.3   Magnitude of electrical betweenness of four successive load buses.

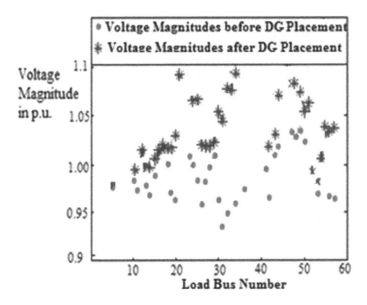

FIGURE 8.4   Voltage magnitude enhancement with implementation of distributed generation.

8.3.3   VULNERABILITY ASSESSMENT USING GVI METRIC INCORPORATING DER: SIMULATION FOR VULNERABILITY ASSESSMENT OF IEEE 57 BUS SYSTEM USING GVI METRIC ON INCORPORATION OF DG

To identify the effective locations to install DG, the power supplying efficiency of each load node needs to be calculated. The determination of shortest path lengths among all of the generators and load nodes is the prerequisite of such calculation. Using the expression of Equation 5.3 (refer to Chapter 5) the power supplying efficiencies of 35 load nodes in the IEEE 57 bus system have been obtained. It has been observed that the ten buses (nodes) having the minimum power supplying efficiencies are the buses numbered as 31, 33, 30, 25, 32, 35, 47, 54, 53, 27 (the power supplying efficiencies of these nodes are 0.1, 0.12, 0.32, 0.36, 0.55, 0.59, 0.63, 0.77, 0.79, and 1.09 respectively). In this simulation, the five load nodes having the lowest power supplying efficiencies have been identified. These load nodes have been numbered as 31, 33, 25, 47, 54 for which the DG installation is recommended. The size of each DG is assumed to be 10MW. With incorporation of new generating buses, the bus numbers are modified as 58~62 and the entire network with (12 generation nodes,35 load nodes and 15 transmitting nodes) 62 nodes has been renumbered. In efficiency-based vulnerability index calculation, e^l denotes the falling ratio of power supplying efficiency. With larger value of e^l the power supplying efficiency declines, hence the related buses or lines will be more vulnerable (Tables 8.1 and 8.2).

The GVI values of the network prior to allocation and post allocation of DGs are shown in Table 8.3.

It has been observed that the GVI values drastically reduce following installation of DG in the designated load buses. Obviously, this reduces the vulnerability of the grid network. In addition to assessment of vulnerability of the system following installation

TABLE 8.1

Vulnerability of Some Nodes of the IEEE 57 Nodes Systems without DG under Attacks on Nodes

Node	e
8(G)	0.3168
12(G)	0.2263
9(G)	0.1983
1(G)	0.1590
11	0.1222
13	0.1090
10	0.0953
43	0.0855
15	0.0830
41	0.0796

TABLE 8.2

Vulnerability of Some Nodes of the IEEE 57 Nodes Systems with DG under Attacks on Nodes

Node	e
8(G)	0.3090
12(G)	0.2165
9(G)	0.1933
1(G)	0.1545
11	0.1166
13	0.1028
10	0.0913
43	0.0831
15	0.0824
41	0.0775
58 (DG)	0.0060
59 (DG)	0.0050
60 (DG)	0.0033
61	0.0006
62	0.0003

of DG, the present simulation includes the effect of DG installed at designated load buses on the system voltage and line loss parameter. Figure 8.5 exhibits the effect of DGs on voltage profiles of the 57 bus system while Table 8.4 indicates the line losses.

The black and grey lines in the Figure 8.5 represents the voltage profiles of buses prior to and post allocation of DGs to designated load buses. It is observed that this methodology of GVI is backed up by improvement of voltage profile of the system. Similarly, the line losses are estimated as shown in the Table 8.4.

TABLE 8.3
GVI Values of IEEE 57 Bus System

Condition	Value of GVI
Before use of DGs	36.5792
Identified load buses(58-62) added with DGs	3.9941

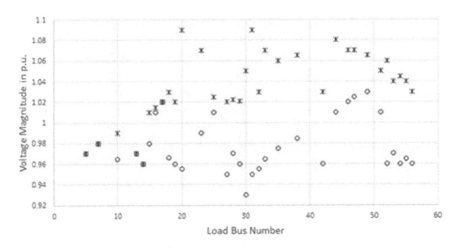

o Voltage Magnitude before DG placement   x Voltage Magnitude after DG placement

FIGURE 8.5   Voltage profile of IEEE 57 bus system with and without DGs.

TABLE 8.4
**Total Active and Reactive Losses in the IEEE 57 Bus
System by Addition of DGs into Designated Load Buses**

Bus No.	Total Loss in MW	Total Loss in MVar
2	27.0318	150.2279
5	25.9881	146.2857
6	22.7915	133.4965
9	20.1473	122.2515
10	19.8262	120.8614
13	18.7594	113.6647
14	18.1836	114.4084
15	17.4407	111.2176
16	15.7057	103.9075
17	15.4488	102.8686
18	15.1619	96.8877

(*Continued*)

TABLE 8.4 (Continued)
Total Active and Reactive Losses in the IEEE 57 Bus
System by Addition of DGs into Designated Load Buses

Bus No.	Total Loss in MW	Total Loss in MVar
19	14.9832	96.3194
20	14.8252	95.9033
23	14.4265	94.4963
25	13.9946	91.7036
27	13.8717	90.9784
28	13.9364	90.8887
29	14.5174	91.4546
30	14.3285	90.3546
31	13.9263	88.8416
32	13.7985	88.4834
33	13.5484	87.8374
35	13.0686	86.2027
38	12.6964	84.7547
41	12.7202	84.0709
42	12.5631	83.2418
43	12.5774	83.2283
44	12.4536	82.6353
47	12.2727	81.8934
49	11.8786	81.0033
50	10.9075	77.8336
51	11.2105	78.8697
52	11.3415	79.6156
53	12.1473	83.2988
54	12.3915	84.1744
55	12.7765	85.6174
56	12.9126	86.2513
57	13.0727	87.0634

The results indicate the gradual declining of line losses with repetitive addition of distribution generation units to load buses. Reduction in the losses is because of the proximity of DG to the load. The employment of large-scale DG units not only decreases the system vulnerability but also achieves better power transmission, increases voltage profile and makes a grid more efficient. In this method, a small change in infrastructure can go a long way in strengthening the power grid.

REFERENCES

1. T. Funabashi, *"Integration of Distributed Energy Resources in Power Systems"*, Cambridge, MA: Academic Press, 2016.
2. hdl.handle.net.
3. D.P. Kothari, K.C. Singal, and R. Ranjan, *"Renewable Energy Sources and Emerging Technologies"*, Second Edition, New Delhi: PHI Learning Private Limited, 2012.

4. G.S. Sawhney, *"Non Conventional Energy Resources"*, New Delhi: PHI Learning Private Limited, 2012.
5. J.P. Barton and D.G. Infield, "Energy storage and its use with intermittent renewable energy," *IEEE Transactions on Energy Conversion*, vol. 19, no. 2, pp. 441–448, 2004. doi: 10.1109/TEC.2003.822305.
6. A.H. Ali, A. Youssef, T. George, and S. Kamel, "Optimal DG allocation in distribution systems using ant lion optimizer,"*2018 International Conference on Innovative Trends in Computer Engineering (ITCE)*, Aswan, pp. 324–331, 2018. doi: 10.1109/ITCE.2018.8316645.
7. K. Mahmoud, N. Yorino, and A. Ahmed, "Optimal distributed generation allocation in distribution systems for loss minimization," *IEEE Transactions on Power Systems*, vol. 31, no. 2, pp. 960–969, 2016. doi: 10.1109/TPWRS.2015.2418333.
8. F.S. Abu-Mouti and M.E. El-Hawary, "Optimal distributed generation allocation and sizing in distribution systems via Artificial Bee Colony algorithm," *IEEE Transactions on Power Delivery*, vol. 26, no. 4, pp. 2090–2101, 2011. doi: 10.1109/TPWRD.2011.2158246.
9. P.J. Varghese, A. Sas, and K. Sunderamoorthy, Optimal siting and sizing of DGs for congestion relief in transmission lines, *2017 IEEE PES Asia-Pacific Power and Energy Engineering Conference (APPEEC)*, Bangalore, 2017.
10. B. Hayes, *"Distributed Generation Systems"*, Amsterdam: Elsevier BV, 2017.
11. T. Chowdhury, A. Chakrabarti, and C.K. Chanda, "Analysis of vulnerability indices of power grid integrated DG units based on Complex Network theory," *Annual IEEE India Conference (INDICON)*, New Delhi, pp. 1–5, 2015.doi: 10.1109/INDICON.2015.7443605.

Appendix A
IEEE 57-BUS TEST SYSTEM (American Electric Power)

Bus Code 1 = Slack Bus, 2 = PV Bus and 0 = PQ Bus

	Bus No.	Bus Code	Voltage Mag.	Angle Degree	Load		Generator				Static Mvar
					MW	Mvar	MW	MVAr	Qmin	Qmax	+Qc/−Ql
Busdata=[	1	1	1.040	0	478	128.9	55	17	0	0	0
	2	2	1.010	−1.18	0	−0.8	3	88	−17	50	0
	3	2	0.985	−5.97	40	−1	41	21	−10	60	0
	4	0	0.981	−7.3	0	0	0	0	0	0	0
	5	0	0.976	−8.52	0	0	13	4	0	0	0
	6	2	0.980	−8.65	0	0.8	75	2	−8	25	0
	7	0	0.984	−7.58	0	0	0	0	0	0	0
	8	2	1.005	−4.45	450	62.1	150	22	−140	200	0
	9	2	0.980	−9.56	0	2.2	121	26	−3	9	0
	10	0	0.986	−11.43	0	0	5	2	0	0	0
	11	0	0.974	−10.17	0	0	0	0	0	0	0
	12	2	1.015	−10.46	310	128.5	337	24	−50	155	0
	13	0	0.979	−9.79	0	0	18	2.3	0	0	0
	14	0	0.970	−9.93	0	0	10.5	5.3	0	0	0
	15	0	0.988	−7.18	0	0	22	5	0	0	0
	16	0	1.013	−8.85	0	0	43	3	0	0	0
	17	0	1.017	−5.39	0	0	42	8	0	0	0
	18	0	1.001	−11.71	0	0	27.2	9.8	0	0	10
	19	0	0.970	−13.20	0	0	3.3	0.6	0	0	0
	20	0	0.964	−13.41	0	0	2.3	1	0	0	0
	21	0	1.008	−12.89	0	0	0	0	0	0	0
	22	0	1.010	−12.84	0	0	0	0	0	0	0
	23	0	1.008	−12.91	0	0	6.3	2.1	0	0	0
	24	0	0.999	−13.25	0	0	0	0	0	0	0
	25	0	0.982	−18.13	0	0	6.3	3.2	0	0	5.9
	26	0	0.959	−12.95	0	0	0	0	0	0	0
	27	0	0.982	−11.48	0	0	9.3	0.5	0	0	0

(*Continued*)

Bus No.	Bus Code	Voltage Mag.	Angle Degree	Load MW	Load Mvar	Generator MW	Generator MVAr	Generator Qmin	Generator Qmax	Static Mvar +Qc/-Ql
28	0	0.997	−10.45	0	0	4.6	2.3	0	0	0
29	0	1.010	−9.75	0	0	17	2.6	0	0	0
30	0	0.962	−18.68	0	0	3.6	1.8	0	0	0
31	0	0.936	−19.34	0	0	5.8	2.9	0	0	0
32	0	0.949	−18.46	0	0	1.6	0.8	0	0	0
33	0	0.947	−18.50	0	0	3.8	1.9	0	0	0
34	0	0.959	−14.10	0	0	0	0	0	0	0
35	0	0.966	−13.86	0	0	6	3	0	0	0
36	0	0.976	−13.59	0	0	0	0	0	0	0
37	0	0.985	−13.41	0	0	0	0	0	0	0
38	0	1.013	−12.71	0	0	14	7	0	0	0
39	0	0.983	−13.46	0	0	0	0	0	0	0
40	0	0.973	−13.62	0	0	0	0	0	0	0
41	0	0.996	−14.05	0	0	6.3	3	0	0	0
42	0	0.966	−15.50	0	0	7.1	4	0	0	0
43	0	1.010	−11.33	0	0	2	1	0	0	0
44	0	1.017	−11.83	0	0	12	1.8	0	0	0
45	0	1.036	−9.25	0	0	0	0	0	0	0
46	0	1.060	−11.09	0	0	0	0	0	0	0
47	0	1.033	−12.49	0	0	29.7	11.6	0	0	0
48	0	1.027	−12.57	0	0	0	0	0	0	0
49	0	1.036	−12.92	0	0	18	8.5	0	0	0
50	0	1.023	−13.39	0	0	21	10.5	0	0	0
51	0	1.052	−12.52	0	0	18	5.3	0	0	0
52	0	0.980	−11.47	0	0	4.9	2.2	0	0	0
53	0	0.971	−12.23	0	0	20	10	0	0	6.3
54	0	0.996	−11.69	0	0	4.1	1.4	0	0	0
55	0	1.031	−10.78	0	0	6.8	3.4	0	0	0
56	0	0.968	−16.04	0	0	7.6	2.2	0	0	0
57	0	0.965	−16.56	0	0	6.7	2.0	0	0	0];

	Bus	Bus	R	X	1/2 B	Line Code = 1 for Lines
	nl	nr	p.u.	p.u.	p.u.	> 1 or < 1 tr. Tap at Bus nl
Linedata=[	1	2	0.0083	0.0280	0.0645	1
	2	3	0.0298	0.0850	0.0409	1
	3	4	0.0112	0.0366	0.0190	1
	4	5	0.0625	0.1320	0.0129	1
	4	6	0.0430	0.1480	0.0174	1
	6	7	0.0200	0.1020	0.0138	1
	6	8	0.0339	0.1730	0.0235	1
	8	9	0.0099	0.0505	0.0274	1
	9	10	0.0369	0.1679	0.0220	1
	9	11	0.0258	0.0848	0.0109	1
	9	12	0.0648	0.2950	0.0386	1
	9	13	0.0481	0.1580	0.0202	1
	13	14	0.0132	0.0434	0.0055	1
	13	15	0.0269	0.0869	0.0115	1
	1	15	0.0178	0.0910	0.0494	1
	1	16	0.0454	0.2060	0.0273	1
	1	17	0.0238	0.1080	0.0143	1
	3	15	0.0162	0.0530	0.0272	1
	4	18	0	0.5550	0	0.97
	4	18	0	0.4300	0	0.978
	5	6	0.0302	0.0641	0.0062	1
	7	8	0.0139	0.0712	0.0097	1
	10	12	0.0277	0.1262	0.0164	1
	11	13	0.0223	0.0732	0.0094	1
	12	13	0.0178	0.0580	0.0302	1
	12	16	0.0180	0.0813	0.0108	1
	12	17	0.0397	0.1790	0.0238	1
	14	15	0.0171	0.0547	0.0074	1
	18	19	0.4610	0.6850	0	1
	19	20	0.2830	0.4340	0	1
	20	21	0	0.7767	0	1.043
	21	22	0.0736	0.1170	0	1
	22	23	0.0099	0.0152	0	1
	23	24	0.1660	0.2560	0.0042	1
	24	25	0	1.1820	0	1
	24	25	0	1.2300	0	1
	24	26	0	0.0473	0	1.043
	26	27	0.1650	0.2540	0	1
	27	28	0.0618	0.0954	0	1
	28	29	0.0418	0.0587	0	1

(Continued)

Bus	Bus	R	X	1/2 B	Line Code = 1 for Lines
nl	nr	p.u.	p.u.	p.u.	> 1 or < 1 tr. Tap at Bus nl
7	29	0	0.0648	0	0.967
25	30	0.1350	0.2020	0	1
30	31	0.3260	0.4970	0	1
31	32	0.507	0.7550	0	1
32	33	0.0392	0.0360	0	1
32	34	0	0.9530	0	0.975
34	35	0.0520	0.0780	0.0016	1
35	36	0.0430	0.0537	0.0008	1
36	37	0.0290	0.0366	0	1
37	38	0.0651	0.1009	0.0010	1
37	39	0.0239	0.0379	0	1
36	40	0.0300	0.0466	0	1
22	38	0.0192	0.0295	0	1
11	41	0	0.7490	0	0.955
41	42	0.2070	0.3520	0	1
41	43	0	0.4120	0	1
38	44	0.0289	0.0585	0.0010	1
15	45	0	0.1042	0	0.955
14	46	0	0.0735	0	0.9
46	47	0.0230	0.0680	0.0016	1
47	48	0.0182	0.0233	0	1
48	49	0.0834	0.1290	0.0024	1
49	50	0.0801	0.1280	0	1
50	51	0.1386	0.2200	0	1
10	51	0	0.0712	0	0.93
13	49	0	0.1910	0	0.895
29	52	0.1442	0.1870	0	1
52	53	0.0762	0.0984	0	1
53	54	0.1878	0.2320	0	1
54	55	0.1732	0.2265	0	1
11	43	0	0.1530	0	0.958
44	45	0.0624	0.1242	0.0020	1
40	56	0	1.1950	0	0.985
56	41	0.5530	0.4590	0	1
56	42	0.2125	0.3540	0	1
39	57	0	1.3550	0	0.98
57	56	0.1740	0.2600	0	1
38	49	0.1150	0.1770	0.0030	1
38	48	0.0312	0.0482	0	1
9	55	0	0.1205	0	0.94];

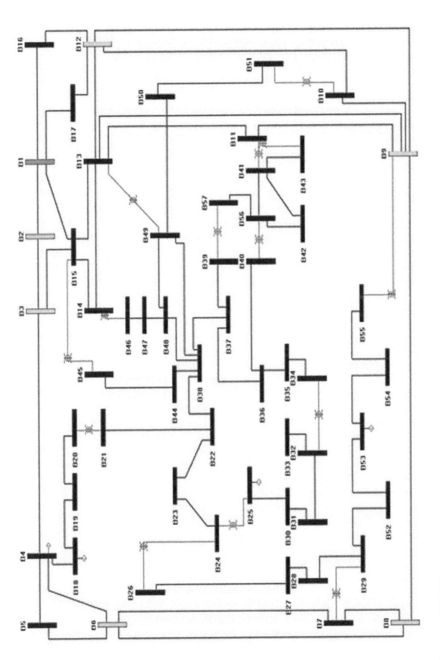

FIGURE A.1 IEEE 57-bus system.

Appendix B
West Bengal State Electricity Board's (WBSEB) 203-Bus System (A Semi-Government Power Utility in Eastern Part of India)

Bus Code 0 = Slack Bus, 2 = PV Bus and 1 = PQ Bus.

Busdata=[

Bus No.	Generator MW	MVAr	Load MW	MVAr	Voltage Mag.	Voltage Angle	Bus Code	ysnt	Qmin	Qmax	
1	4	0	0	0	1.05	0	0	0	−1	2;	%FKK21
2	1.8	0	0	0	1.04	0	2	0	−0.5	1;	%KT115.75
3	1.8	0	0	0	1.04	0	2	0	−0.5	1;	%KT215.75
4	1.9	0	0	0	1.04	0	2	0	−0.5	1;	%BKR15.75
5	1.8	0	0	0	1.05	0	2	0	−0.5	1;	%FKK15.75
6	1.9	0	0	0	1.0433	0	2	0	−0.5	1;	%BAKR15.7
7	0.6	0	0	0	1.02	0	2	0	−0.3	0.6;	%STP13.8
8	0.9	0	0	0	1	0	2	0	−0.22	0.45;	%DPL13.8
9	0.115	0	0	0	1.02	0	2	0	−0.03	0.06;%RMM211	
10	1.8	0	0	0	1	0	2	0	−0.5	1;	%BTP11
11	0.5	0	0	0	1	0	2	0	−0.2	0.4;	%BTPS11
12	0.6	0	0	0	1	0	2	0	−0.15	0.3;	%DPL11
13	0.675	0	0	0	1.01	0	2	0	−0.2	0.4;	%CHU11
14	0.181	0	0	0	1.02	0	2	0	−0.02	0.04;	%JAL111
15	0.136	0	0	0	1.02	0	2	0	−0.009	0.018;	%JAL211
16	0.18	0	0	0	1.02	0	2	0	−0.045	0.09;	%RANGIT11
17	1.5	0	0	0	1.04	0	2	0	−0.45	0.9;	%MEJIA11
18	0.5	0	0	0	1.03	0	2	0	−0.3	0.6;	%WARIA11
19	1.5	0	0	0	1.03	0	2	0	−0.5	1;	%WARIA211
20	1.5	0	0	0	1.04	0	2	0	−0.4	0.8;	%TALA11
21	0.068	0	0	0	1.02	0	2	0	−0.019	0.038;	%TCF16.6
22	0.075	0	0	0	1.02	0	2	0	−0.019	0.038;	%TCF26.6
23	0.075	0	0	0	1.02	0	2	0	−0.019	0.038;	%TCF36.6

(Continued)

Bus No.	Generator MW	MVAr	Load MW	MVAr	Voltage Mag.	Voltage Angle	Bus Code	ysnt	Qmin	Qmax	
24	0.2	0	0	0	1	0	2	0	-0.05	0.1;	%DPL6.3
25	0	0	0	0	1	0	1	0	0	0;	%PSP16.5
26	0	0	1.2	1	1	0	1	0	0	0;	%JRT132
27	0	0	0.68	0.921	1	0	1	0	0	0;	%ARAM132
28	0	0	0	0	1	0	1	0	0	0;	%KTPS132
29	0	0	0	0	1	0	1	0	0	0;	%MALPG132
30	0	0	0	0	1	0	1	0	0	0;	%DRG132
31	0	0	0.55	0.34	1	0	1	0	0	0;	%DOMJ132
32	0	0	0.51	0.316	1	0	1	0.067	0	0;	%GOK132
33	0	0	0.5	1.32	1	0	1	0.067	0	0;	%HOW132
34	0	0	0.6	1.363	1	0	1	0	0	0;	%KASBA132
35	0	0	0.51	0.316	1	0	1	0	0	0;	%LXP132
36	0	0	0	0	1	0	1	0	0	0;	%MIDNA132
37	0	0	0.35	0.217	1	0	1	0	0	0;	%NJP132
38	0	0	0.2	0.124	1	0	1	0	0	0;	%NHAL132
39	0	0	0	0	1	0	1	0	0	0;	%RISH132
40	0	0	0.45	0.527	1	0	1	0.067	0	0;	%SATG132
41	0	0	1	1.239	1	0	1	0	0	0;	%DPL132
42	0	0	0	0	1	0	1	0	0	0;	%BRPPG132
43	0	0	0	0	1	0	1	0	0	0;	%SLGPG132
44	0	0	0.72	0.446	1	0	1	0.033	0	0;	%ADI132
45	0	0	0.15	0.193	1	0	1	0	0	0;	%ALIP132

(Continued)

Bus No.	Generator MW	Generator MVAr	Load MW	Load MVAr	Voltage Mag.	Voltage Angle	Bus Code	ysnt	Qmin	Qmax	
46	0	0	0.41	0.178	1	0	1	0	0	0;	%ASOK132
47	0	0	0.32	0.198	1	0	1	0	0	0;	%BALU132
48	0	0	0.58	0.259	1	0	1	0.033	0	0;	%BNK132
49	0	0	0.37	0.477	1	0	1	0.033	0	0;	%BARAS132
50	0	0	0.35	0.2169	1	0	1	0	0	0;	%BASIR132
51	0	0	0.4	0.2558	1	0	1	0.0333	0	0;	%JOKA132
52	0	0	0.46	0.533	1	0	1	0.0333	0	0;	%BERH132
53	0	0	0.35	0.2169	1	0	1	0	0	0;	%BRP132
54	0	0	0.4	0.2958	1	0	1	0	0	0;	%VISH132
55	0	0	0	0	1	0	1	0	0	0;	%BOLP132
56	0	0	0.33	0.2045	1	0	1	0	0	0;	%BONG132
57	0	0	0.4	0.2479	1	0	1	0	0	0;	%CLC132
58	0	0	0.53	0.3285	1	0	1	0.333	0	0;	%CHK132
59	0	0	0.14	0.0868	1	0	1	0	0	0;	%DALK132
60	0	0	0.24	0.1487	1	0	1	0	0	0;	%DARJ132
61	0	0	0.31	0.1921	1	0	1	0.167	0	0;	%DEBOI132
62	0	0	0.57	0.3533	1	0	1	0.15	0	0;	%DHRAM132
63	0	0	0.25	0.1549	1	0	1	0	0	0;	%DHUL132
64	0	0	0	0	1	0	1	0	0	0;	%EGRA132
65	0	0	0.43	0.2665	1	0	1	0.1667	0	0;	%FALTA132
66	0	0	0.39	0.2417	1	0	1	0	0	0;	%HALD132
67	0	0	0.49	0.4276	1	0	1	0	0	0;	%HIZLI132

(Continued)

Bus No.	Generator MW	Generator MVAr	Load MW	Load MVAr	Voltage Mag.	Voltage Angle	Bus Code	ysnt	Qmin	Qmax	
68	0	0	0.66	0.409	1	0	1	0.1333	0	0;	%KLY132
69	0	0	0.74	0.4586	1	0	1	0.1667	0	0;	%KATWA132
70	0	0	0.41	0.2541	1	0	1	0	0	0;	%KHAN132
71	0	0	0.46	0.5702	1	0	1	0.1333	0	0;	%KOLAG132
72	0	0	0.52	0.3223	1	0	1	0.196	0	0;	%KRISH132
73	0	0	0.56	0.3325	1	0	1	0.1667	0	0;	%LILO132
74	0	0	0.68	0.4214	1	0	1	0	0	0;	%MALDA132
75	0	0	0.33	0.2045	1	0	1	0	0	0;	%MANK132
76	0	0	0.4	0.2479	1	0	1	0	0	0;	%MOINA132
77	0	0	0	0	1	0	1	0	0	0;	%NBU132
78	0	0	0.25	0.3409	1	0	1	0	0	0;	%PING132
79	0	0	0.45	0.4648	1	0	1	0	0	0;	%PURU132
80	0	0	0.46	0.2851	1	0	1	0.3333	0	0;	%RGJ132
81	0	0	0.23	0.1425	1	0	1	0	0	0;	%RAIG132
82	0	0	0.56	0.3471	1	0	1	0	0	0;	%RAINA132
83	0	0	0.4	0.2479	1	0	1	0	0	0;	%RAMP132
84	0	0	0.7	0.6197	1	0	1	0	0	0;	%RANA132
85	0	0	0.5	0.3099	1	0	1	0.1667	0	0;	%SAIN132
86	0	0	0	0	1	0	1	0	0	0;	%SL132
87	0	0	0.32	0.1983	1	0	1	0.0333	0	0;	%SAMSI132
88	0	0	0.42	0.2603	1	0	1	0	0	0;	%SLG132
89	0	0	0.25	0.2048	1	0	1	0	0	0;	%SONA132

(Continued)

Bus No.	Generator MW	Generator MVAr	Load MW	Load MVAr	Voltage Mag.	Voltage Angle	Bus Code	ysnt	Qmin	Qmax	
90	0	0	0.29	0.1797	1	0	1	0	0	0;	%TAML132
91	0	0	0.54	0.3347	1	0	1	0	0	0;	%TARAK132
92	0	0	0	0	1	0	1	0	0	0;	%TITA132
93	0	0	0.5	0.3099	1	0	1	0	0	0;	%UKHRA132
94	0	0	0.57	0.3533	1	0	1	0	0	0;	%ULU132
95	0	0	0	0	1	0	1	0	0	0;	%STPS132
96	0	0	0	0	1	0	1	0	0	0;	%BTPS132
97	0	0	0.19	0.1178	1	0	1	0	0	0;	%BALICI132
98	0	0	0	0	1	0	1	0	0	0;	%TCFI132
99	0	0	0	0	1	0	1	0	0	0;	%TCFII132
100	0	0	0	0	1	0	1	0	0	0;	%TCF3132
101	0	0	0	0	1	0	1	0	0	0;	%RMM2132
102	0	0	0.13	0.0806	1	0	1	0	0	0;	%HM132
103	0	0	0.05	0.031	1	0	1	0	0	0;	%DANK132
104	0	0	0	0	1	0	1	0	0	0;	%RANGT132
105	0	0	0.1	0.062	1	0	1	0	0	0;	%HPCL132
106	0	0	0.006	0.0037	1	0	1	0	0	0;	%FCI132
107	0	0	0.23	0.1425	1	0	1	0	0	0;	%JANGI132
108	0	0	0.31	0.1921	1	0	1	0	0	0;	%CTOLA132
109	0	0	0.12	0.0744	1	0	1	0.3096	0	0;	%GRAM132
110	0	0	0.35	0.2169	1	0	1	0	0	0;	%GTOK132
111	0	0	0.28	0.1735	1	0	1	0	0	0;	%MELLI132

(*Continued*)

Bus No.	Generator		Load		Voltage Mag.	Voltage Angle	Bus Code	ysnt	Qmin	Qmax	
	MW	MVAr	MW	MVAr							
112	0	0	1	1.2395	1	0	1	0	0	0;	%WARIA132
113	0	0	0.38	0.2355	1	0	1	0	0	0;	%BIRSG132
114	0	0	0.35	0.2169	1	0	1	0	0	0;	%COOCH132
115	0	0	0	0	1	0	1	0	0	0;	%JRT220
116	0	0	0	0	1	0	1	0	0	0;	%ARAM220
117	0	0	0	0	1	0	1	0	0	0;	%KTPS220
118	0	0	0	0	1	0	1	0	0	0;	%BAKR220
119	0	0	0	0	1	0	1	0	0	0;	%PARU220
120	0	0	0	0	1	0	1	0	0	0;	%MALPG220
121	0	0	0	0	1	0	1	0	0	0;	%BGURI220
122	0	0	0	0	1	0	1	0	0	0;	%DRG220
123	0	0	0	0	1	0	1	0	0	0;	%DOMJ220
124	0	0	0	0	1	0	1	0	0	0;	%GOK220
125	0	0	0	0	1	0	1	0	0	0;	%HOW220
126	0	0	0	0	1	0	1	0	0	0;	%KASBA220
127	0	0	0	0	1	0	1	0	0	0;	%LXP220
128	0	0	0	0	1	0	1	0	0	0;	%MIDNA220
129	0	0	0	0	1	0	1	0	0	0;	%NJP220
130	0	0	0	0	1	0	1	0	0	0;	%NHAL220
131	0	0	0	0	1	0	1	0	0	0;	%RISH220
132	0	0	0	0	1	0	1	0	0	0;	%SATG220
133	0	0	0	0	1	0	1	0	0	0;	%DPL220

(Continued)

Bus No.	Generator		Load		Voltage Mag.	Voltage Angle	Bus Code	ysnt	Qmin	Qmax	
	MW	MVAr	MW	MVAr							
134	0	0	0	0	1	0	1	0	0	0;	%WARIA220
135	0	0	0	0	1	0	1	0	0	0;	%CHUK220
136	0	0	0.3	0.1859	1	1	0	0	0	0;	%BRPPG220
137	0	0	0	0	1	0	1	0	0	0;	%SLGPG220
138	0	0	0.5	0.5099	1	1	0	0	0	0;	%DALPG220
139	0	0	0	0	1	0	1	0	0	0;	%PURN220
140	0	0	0.5	0.6197	1	1	0	0	0	0;	%STPS220
141	0	0	1	2.479	1	0	1	0	0	0;	%MEJIA220
142	0	0	0.45	0.4789	1	1	0	0	0	0;	%FKK220
143	0	0	0	0	1	0	1	0	0	0;	%SGPG220
144	0	0	0	0	1	0	1	0	0	0;	%KRIS220
145	0	0	1	0.35	1	0	1	-0.5333	0	0;	%JRT400 load added
146	0	0	1	1	1	0	1	-0.5667	0	0;	%ARAM400load added
147	0	0	0	0	1	0	1	-0.5867	0	0;	%KTPS400
148	0	0	0	0	1	0	1	-0.5	0	0;	%BAKR400
149	0	0	2	3.479	1	0	1	-0.5333	0	0;	%FKK400
150	0	0	3.5	3.8592	1	0	1	-0.5	0	0;	%PARU400
151	0	0	0	0	1	0	1	-0.42	0	0;	%MALPG400
152	0	0	0	0	1	0	1	-0.41	0	0;	%BGURI400
153	0	0	0.5	0.6197	1	0	1	-0.525	0	0;	%BARIP400
154	0	0	0.75	1.8592	1	0	1	-0.5867	0	0;	%PURN400

(Continued)

Bus No.	Generator MW	MVAr	Load MW	MVAr	Voltage Mag.	Voltage Angle	Bus Code	ysnt	Qmin	Qmax	
155	0	0	0	0	1	0	1	-0.5366	0	0;	%SGPG400
156	0	0	0	0	1	0	1	-0.5366	0	0;	%TALA400
157	0	0	0	0	1	0	1	-0.5366	0	0;	%PPSP2
158	0	0	0	0	1	0	1	0	0	0;	%JALI66
159	0	0	0	0	1	0	1	0	0	0;	%JALII66
160	0	0	0.12	0.0744	1	0	1	0.383	0	0;	%CHALSA66 cap added
161	0	0	0.16	0.0992	1	0	1	0.433	0	0;	%BANAR66
162	0	0	0	0	1	0	1	0.583	0	0;	%BRP66
163	0	0	0.15	0.093	1	0	1	0.583	0	0;	%KPONG66
164	0	0	0.23	0.018	1	0	1	0.613	0	0;	%MELLI66
165	0	0	0.2	0.32	1	0	1	0	0	0;	%JRT33
166	0	0	0.1	0.21	1	0	1	0	0	0;	%ARAM33
167	0	0	0.23	0.15	1	0	1	0	0	0;	%KTPS33
168	0	0	0.31	0.21	1	0	1	0	0	0;	%FKK33
169	0	0	0.1	0.21	1	0	1	0	0	0;	%PARU33
170	0	0	0.1	0.21	1	0	1	0	0	0;	%BAKR33
171	0	0	0.1	0.2	1	0	1	0	0	0;	%MLDAPG33
172	0	0	0.1	0.35	1	0	1	0	0	0;	%BGURI33
173	0	0	0.23	0.41	1	0	1	0	0	0;	%PURN33
174	0	0	0.1	0.25	1	0	1	0.05	0	0;	%JRT133
175	0	0	0.1	0.21	1	0	1	0	0	0;	%ARAM133

(*Continued*)

Bus No.	Generator MW	Generator MVAr	Load MW	Load MVAr	Voltage Mag.	Voltage Angle	Bus Code	ysnt	Qmin	Qmax	
176	0	0	0.1	0.21	1	0	1	0	0	0;	%KTPS133
177	0	0	0.1	0.45	1	0	1	0	0	0;	%MLDPG133
178	0	0	0.1	0.21	1	0	1	0	0	0;	%DRG33
179	0	0	0.1	0.21	1	0	1	0	0	0;	%DOMJ33
180	0	0	0.1	0.21	1	0	1	0	0	0;	%GOK33
181	0	0	0.1	0.21	1	0	1	0	0	0;	%HOW33
182	0	0	0.1	0.21	1	0	1	0	0	0;	%KASBA33
183	0	0	0.1	0.21	1	0	1	0	0	0;	%LXP33
184	0	0	0.1	0.21	1	0	1	0	0	0;	%MIDNA33
185	0	0	0.1	0.51	1	0	1	0	0	0;	%NJP33
186	0	0	0.1	0.21	1	0	1	0	0	0;	%NHAL33
187	0	0	0.1	0.21	1	0	1	0	0	0;	%RISH33
188	0	0	0.1	0.1	1	0	1	0	0	0;	%SATG33
189	0	0	0.1	0.1	1	0	1	0	0	0;	%DPL33
190	0	0	0.1	0.4	1	0	1	0	0	0;	%BRPPG33
191	0	0	0.1	0.4	1	0	1	0	0	0;	%SLGPG33
192	0	0	0.1	0.1	1	0	1	0	0	0;	%STPS33
193	0	0	0.1	0.1	1	0	1	0	0	0;	%WARIA33
194	0	0	0.37	0.2176	1	0	1	0.38	0	0;	%RISHRA33
195	0	0	0.36	0.018	1	0	1	0.5233	0	0;	%BISHNU33
196	0	0	0.17	0.1413	1	0	1	0.5233	0	0;	%EGRA33
197	0	0	0.41	0.2161	1	0	1	0.03	0	0;	%NBU33

(Continued)

Bus No.	Generator		Load		Voltage Mag.	Voltage Angle	Bus Code	ysnt	Qmin	Qmax	
	MW	MVAr	MW	MVAr							
198	0	0	0.21	0.12	1	0	1	0.5367	0	0;	%SLAKE33
199	0	0	0.23	0.1599	1	0	1	0.372	0	0;	%BOLPUR33
200	0	0	0.25	0.30	1	0	1	0.5883	0	0;	%TITAG33
201	0	0	0.32	0.2	1	0	1	0.5833	0	0;	%MIDNAP33
202	0	0	0.1	0.25	1	0	1	0.15	0	0;	%SGPG33
203	0	0	0	0	1	0	1	0	0	0;]	%KRIS33

Actual tap-position and tap of trf. is included.

Slack bus should be from bus for line 206,

Data with shunt capacitor only.

tap_pos=1 for tr at SE and tap_pos=2 for tr at RE and zero for tr. Line.

	Bus No.	fb	eb	r	x	hlch	Tap	tap posl		lmt
Linedata=[	52	44	96	0.0036	0.0071	0.0108	1.0	0	10;	%ADI132-BTPS132
	53	44	91	0.0095	0.0189	0.0287	1.0	0	10;	%ADI132-TARAK132
	54	45	53	0.0385	0.077	0.0292	1.0	0	10;	%ALIP132-BRP132
	55	27	82	0.0116	0.0232	0.0352	1.0	0	10;	%ARAM132-RAINA132
	56	27	91	0.0071	0.0143	0.0217	1.0	0	10;	%ARAM132-TARAK132
	57	46	57	0.031	0.0621	0.0235	1.0	0	10;	%ASOK132-CLC132
	58	46	26	0.0053	0.0107	0.0162	1.0	0	10;	%ASOK132-JRT132
	59	36	97	0.0178	0.0357	0.0135	1.0	0	10;	%MIDNA132-BALIC132
	60	78	97	0.0185	0.0371	0.0141	1.0	0	10;	%PING132-BALIC132
	61	47	109	0.0407	0.0813	0.0309	1.0	0	10;	%BALU132-GRAM132
	62	48	54	0.0134	0.0268	0.0407	1.0	0	10;	%BNK132-VISH132
	63	49	50	0.0125	0.025	0.0379	1.0	0	10;	%BARAS132-BASIR132
	64	49	26	0.0078	0.0157	0.0238	1.0	0	10;	%BARAS132-JRT132
	65	52	32	0.0064	0.0128	0.0195	1.0	0	10;	%BERH132-GOK132
	66	53	42	0.0006	0.0011	0.0004	1.0	0	10;	%BRP132-BRPPG132
	67	53	76	0.0132	0.0264	0.0401	1.0	0	10;	%BRP132-MOINA132
	68	55	30	0.0481	0.0961	0.0365	1.0	0	10;	%BOLP132-DRG132
	69	55	85	0.026	0.0519	0.0197	1.0	0	10;	%BOLP132-SAIN132
	70	56	72	0.0189	0.0377	0.0573	1.0	0	10;	%BONG132-KRISH132
	71	96	62	0.0073	0.0146	0.0222	1.0	0	10;	%BTPS132-DHRAM132
	72	96	62	0.0188	0.0377	0.0143	1.0	0	10;	%BTPS132-DHRAM132
	73	96	70	0.0285	0.0571	0.0217	1.0	0	10;	%BTPS132-KHAN132
	74	96	68	0.0107	0.0214	0.0081	1.0	0	10;	%BTPS132-KLY132
	75	96	73	0.0328	0.0656	0.0249	1.0	0	10;	%BTPS132-LILO132
	76	96	39	0.0111	0.0221	0.0336	1.0	0	10;	%BTPS132-RISH132
	77	96	40	0.0378	0.0756	0.0287	1.0	0	10;	%BTPS132-SATG132
	78	96	108	0.0189	0.0378	0.0143	1.0	0	10;	%BTPS132-CTOLA132
	79	58	36	0.0357	0.0713	0.0271	1.0	0	10;	%CHK132-MIDNA132

(Continued)

Bus No.	fb	eb	r	x	hlch	Tap	tap posl	lmt
80	58	54	0.0321	0.0642	0.0244	1.0	0	10; %CHK132-VISH132
81	59	81	0.0182	0.0364	0.0552	1.0	0	10; %DALK132-RAIG132
82	59	100	0.0388	0.0776	0.1178	1.0	0	10; %DALK132-TCF3132
83	60	77	0.0467	0.0934	0.0355	1.0	0	10; %DARJ132-NBU132
84	60	101	0.0146	0.0292	0.0111	1.0	0	0	10; %DARJ132-RMM2132
85	61	69	0.009	0.018	0.0274	1.0	0	10; %DEBO132-KATWA132
86	62	26	0.005	0.01	0.0152	1.0	0	10; %DHRAM132-JRT132
87	62	68	0.0093	0.0185	0.007	1.0	0	10; %DHRAM132-KLY132
88	62	84	0.0132	0.0264	0.0401	1.0	0	10; %DHRAM132-RANA132
89	62	92	0.0103	0.0207	0.0314	1.0	0	10; %DHRAM132-TITA132
90	63	74	0.0541	0.1081	0.041	1.0	0	10; %DHUL132-MALDA132
91	63	32	0.0499	0.0999	0.0379	1.0	0	10; %DHUL132-GOK132
92	31	94	0.0096	0.0193	0.0292	1.0	0	10; %DOMJ132-ULU132
93	30	41	0.0036	0.0071	0.0108	1.0	0	10; %DRG132-DPL132
94	30	75	0.0139	0.0278	0.0422	1.0	0	10; %DRG132-MANK132
95	30	85	0.0421	0.0842	0.0319	1.0	0	10; %DRG132-SAIN132
96	30	93	0.0065	0.013	0.0197	1.0	0	10; %DRG132-UKHRA132
97	30	54	0.0185	0.0371	0.0563	1.0	0	10; %DRG132-VISH132
98	64	67	0.0435	0.087	0.033	1.0	0	10; %EGRA132-HIZLI132
99	65	51	0.0206	0.0412	0.0156	1.0	0	10; %FALTA132-JOKA132
100	65	35	0.0121	0.0243	0.0368	1.0	0	10; %FALTA132-LXP132
101	51	89	0.0128	0.0257	0.0097	1.0	0	10; %JOKA132-SONA132
102	32	69	0.0203	0.0407	0.0617	1.0	0	10; %GOK132-KATWA132
103	32	80	0.0378	0.0756	0.0287	1.0	0	10; %GOK132-RGJ132
104	32	83	0.0414	0.0827	0.0314	1.0	0	10; %GOK132-RAMP132
105	32	85	0.0385	0.077	0.0292	1.0	0	10; %GOK132-SAIN132
106	66	90	0.0143	0.0285	0.0433	1.0	0	10; %HALD132-TAML132

(Continued)

Bus No.	fb	eb	r	x	hlch	Tap	tap posl	lmt
107	102	103	0.0043	0.0086	0.0032	1.0	0	10; %HM132-DANK132
108	103	73	0.0043	0.0086	0.0032	1.0	0	10; %DANK132-LILO132
109	102	39	0.0078	0.0157	0.006	1.0	0	10; %HM132-RISH132
110	67	36	0.0053	0.0107	0.0162	1.0	0	10; %HIZLI132-MIDNA132
111	33	73	0.0043	0.0086	0.013	1.0	0	10; %HOW132-LILO132
112	33	73	0.0046	0.0093	0.0141	1.0	0	10; %HOW132-LILO132
113	26	56	0.0126	0.0251	0.0381	1.0	0	10; %JRT132-BONG132
114	51	34	0.0208	0.0417	0.0158	1.0	0	10; %JOKA132-KASBA132
115	51	35	0.0272	0.0545	0.0207	1.0	0	10; %JOKA132-LXP132
116	34	86	0.009	0.0179	0.0272	1.0	0	10; %KASBA132-SL132
117	34	89	0.0087	0.0174	0.0066	1.0	0	10; %KASBA132-SONA132
118	69	40	0.0155	0.031	0.0471	1.0	0	10; %KATWA132-SATG132
119	70	40	0.01	0.02	0.0076	1.0	0	10; %KHAN132-SATG132
120	28	71	0.0012	0.0025	0.0038	1.0	0	10; %KTPS132-KOLAG132
121	28	90	0.0089	0.0178	0.0271	1.0	0	10; %KTPS132-TAML132
122	28	94	0.0104	0.0208	0.0315	1.0	0	10; %KTPS132-ULU132
123	73	39	0.0136	0.0271	0.0103	1.0	0	10; %LILO132-RISH132
124	73	39	0.0136	0.0271	0.0103	1.0	0	10; %LILO132-RISH132
125	74	29	0.0021	0.0041	0.0063	1.0	0	10; %MALDA132-MALPG132
126	74	81	0.0556	0.1113	0.0422	1.0	0	10; %MALDA132-1RAIG132
127	74	87	0.0349	0.0699	0.0265	1.0	0	10; %MALDA132-SAMSI132
128	36	78	0.0285	0.0571	0.0217	1.0	0	10; %MIDNA132-PING132
129	36	54	0.0678	0.1355	0.0514	1.0	0	10; %MIDNA132-VISH132
130	37	76	0.02	0.0399	0.0606	1.0	0	10; %NJP132-MOINA132
131	77	101	0.0492	0.0984	0.0374	1.0	0	10; %NBU132-RMM2132
132	77	43	0.0071	0.0143	0.0054	1.0	0	10; %NBU132-SLGPG132
133	77	98	0.0142	0.0284	0.0108	1.0	0	10; %NBU132-TCFI132
134	37	98	0.0143	0.0285	0.0108	1.0	0	10; %NJP132-TCFI132

(Continued)

Bus No.	fb	eb	r	x	hlch	Tap	tap posl	lmt
135	37	88	0.0046	0.0093	0.0141	1.0	0	10; %NJP132-SLG132
136	95	79	0.0125	0.025	0.0379	1.0	0	10; %STPS132-PURU132
137	87	81	0.0506	0.1013	0.0384	1.0	0	10; %SAMSI132-RAIG132
138	85	83	0.0314	0.0628	0.0238	1.0	0	10; %SAIN132-RAMP132
139	40	75	0.0257	0.0514	0.078	1.0	0	10; %SATG132-MANK132
140	98	99	0.0056	0.0113	0.0043	1.0	0	10; %TCFI132-TCFII132
141	98	100	0.0178	0.0357	0.0135	1.0	0	10; %TCFI132-TCF3132
142	99	100	0.0121	0.0243	0.0092	1.0	0	10; %TCFII132-TCF3132
143	77	37	0.0071	0.0143	0.0054	1.0	0	10; %NBU132-NJP132
144	43	104	0.0317	0.0635	0.0964	1.0	0	10; %SLGPG132-RANGT132
145	104	101	0.0193	0.0385	0.0146	1.0	0	10; %RANGT132-RMM2132
146	38	66	0.0014	0.0029	0.0011	1.0	0	10; %NHAL132-HALD132
147	38	105	0.0007	0.0014	0.0005	1.0	0	10; %NHAL132-HPCL132
148	66	105	0.0014	0.0029	0.0011	1.0	0	10; %HALD132-HPCL132
149	39	108	0.0039	0.0078	0.003	1.0	0	10; %RISH132-CTOLA132
150	31	107	0.0061	0.0121	0.0184	1.0	0	10; %DOMJ132-JANGI132
151	66	106	0.0018	0.0036	0.0054	1.0	0	10; %HALD132-FCI132
152	81	109	0.0407	0.0813	0.0309	1.0	0	10; %RAIG132-GRAM132
153	27	113	0.0082	0.0164	0.0249	1.0	0	10; %ARAM132-BIRSG132
154	104	110	0.0521	0.1041	0.0395	1.0	0	10; %RANGT132-GTOK132
155	104	111	0.0214	0.0428	0.0162	1.0	0	10; %RANGT132-MELLI132
156	110	111	0.0428	0.0856	0.0325	1.0	0	10; %GTOK132-MELLI132
157	111	43	0.0642	0.1284	0.0487	1.0	0	10; %MELLI132-SLGPG132
158	104	43	0.0663	0.1327	0.0503	1.0	0	10; %RANGT132-SLGPG132
159	114	45	0.0143	0.0285	0.0108	1.0	0	10; %COOCH132-ALIP132

(*Continued*)

Bus No.	fb	eb	r	x	hlch	Tap	tap posl		lmt
160	114	53	0.0514	0.1027	0.039	1.0	0	10;	%COOCH132-BRP132
22	116	122	0.0166	0.058	0.2425	1.0	0	10;	%ARAM220-DRG220
23	122	119	0.0014	0.0049	0.0207	1.0	0	10;	%DRG220-PARU220
24	116	123	0.0037	0.013	0.2169	1.0	0	10;	%ARAM220-DOMJ220
25	116	128	0.0046	0.016	0.2673	1.0	0	10;	%ARAM220-MIDNA220
26	116	131	0.0094	0.0328	0.1372	1.0	0	10;	%ARAM220-RISH220
27	116	140	0.0125	0.0438	0.7335	1.0	0	10;	%ARAM220-STPS220
28	118	122	0.0025	0.0088	0.1477	1.0	0	10;	%BAKR220-2DRG220
29	118	124	0.0053	0.0184	0.3083	1.0	0	10;	%BAKR220-GOK220
30	118	132	0.009	0.0315	0.5263	1.0	0	10;	%BAKR220-SATG220
31	123	125	0.0012	0.0041	0.0684	1.0	0	10;	%DOMJ220-HOW220
32	122	133	0.0006	0.0022	0.0376	1.0	0	10;	%DRG220-DPL220
33	122	140	0.0064	0.0225	0.3759	1.0	0	10;	%DRG220-STPS220
34	125	117	0.0046	0.016	0.2669	1.0	0	10;	%HOW220-KTPS220
35	115	126	0.0038	0.0133	0.2233	1.0	0	10;	%JRT220-KASBA220
36	115	126	0.0076	0.0267	0.1118	1.0	0	10;	%JRT220-KASBA220
37	115	132	0.0051	0.0177	0.297	1.0	0	10;	%JRT220-SATG220
38	126	127	0.0082	0.0287	0.1201	1.0	0	10;	%KASBA220-LXP220
39	117	130	0.0036	0.0126	0.2109	1.0	0	10;	%KTPS220-NHAL220
40	119	134	0.001	0.0036	0.0602	1.0	0	10;	%PARU220-WARIA220
41	122	134	0.0011	0.0038	0.0639	1.0	0	10;	%DRG220-WARIA220
42	136	135	0.0055	0.0192	0.7218	1.0	0	10;	%BRPPG220-CHUK220
43	138	120	0.0069	0.0243	0.406	1.0	0	10;	%DALPG220-MALPG220
44	138	137	0.0076	0.0265	0.4436	1.0	0	10;	%DALPG220-SLGPG220
45	121	136	0.0053	0.0186	0.312	1.0	0	10;	%BGURI220-BRPPG220

(Continued)

Bus No.	fb	eb	r	x	hlch	Tap	tap posl	lmt
46	121	137	0.0001	0.0004	0.0075	1.0	0	10; %BGURI220-SLGPG220
48	141	134	0.0022	0.0076	0.1278	1.0	0	10; %141MEJIA220-WARIA220
49	132	144	0.003	0.0106	0.1767	1.0	0	10; %SATG220-KRIS220
50	143	115	0.0091	0.0319	0.1335	1.0	0	10; %SGPG220-JRT220
51	127	143	0.0055	0.0193	0.0808	1.0	0	10; %LXP220-SGPG220
7	146	148	0.004	0.0151	0.9204	1.0	0	10; %ARAM400-8BAKR400
8	145	148	0.005	0.0189	1.154	1.0	0	10; %JRT400-BAKR400
9	146	147	0.002	0.0075	0.4545	1.0	0	10; %ARAM400-KTPS400
10	145	147	0.0042	0.0156	0.9516	1.0	0	10; %JRT400-KTPS400
11	147	153	0.0056	0.021	1.2783	1.0	0	10; %KTPS400-BARIP400
12	145	149	0.0073	0.0273	1.6618	1.0	0	10; %JRT400-FKK400
13	149	150	0.0023	0.0087	2.1163	1.0	0	10; %FKK400-PARU400
14	149	151	0.0006	0.0022	0.5255	1.0	0	10; %FKK400-MALPG400
15	151	154	0.0026	0.0099	2.4145	1.0	0	10; %MALPG400-PURN400
16	152	154	0.0025	0.0093	2.2725	1.0	0	10; %BGURI400-PURN400
17	150	157	0.0061	0.0228	1.3919	1.0	0	10; %PARU400-PPSP2
18	146	157	0.0032	0.0122	2.9685	1.0	0	10; %ARAM400-PPSP2
19	155	149	0.0095	0.0355	2.166	1.0	0	10; %SGPG400-FKK400
20	155	145	0.0022	0.0083	0.5042	1.0	0	10; %SGPG400-JRT400
21	152	156	0.0034	0.0128	3.1247	1.0	0	10; %BGURI400-TALA400
1	158	159	0.0114	0.0128	0.0009	1.0	0	10; %JALI66-JALII66
2	159	160	0.1027	0.1155	0.0081	1.0	0	10; %JALII66-CHALSA66
3	160	161	0.1912	0.215	0.0038	1.0	0	10; %CHALSA66-BANAR66
4	161	162	0.1096	0.1233	0.0022	1.0	0	10; %BANAR66-BRP66
5	160	163	0.131	0.1723	0.0065	1.0	0	10; %CHALSA66-KPONG66 Zlinechng
6	164	163	0.0571	0.0642	0.0011	1.0	0	10; %MELLI66-KPONG66
161	145	115	0	0.0132	0	0.95	1	10; %JRT400-JRT220 tap chngori 1.05
162	146	116	0	0.0132	0	0.95	1	10; %ARAM400-ARAM220 tap chng ori 1.05

(Continued)

Bus No.	fb	eb	r	x	hlch	Tap	tap posl	lmt
163	147	117	0	0.0198	0	1.0	1	10; %KTPS400-KTPS220
164	148	118	0	0.0198	0	1.0	1	10; %BAKR400-BAKR220
165	150	119	0	0.0198	0	1.05	1	10; %PARU400-PARU220 tap chng ori 1.05
166	151	120	0	0.0198	0	1.0	1	10; %MALPG400-MALPG220
167	152	121	0	0.0198	0	1	1	10; %BGURI400-BGURI220 tap chng ori −1
168	154	139	0	0.0198	0	1	1	10; %PURN400-PURN220
169	149	142	0	0.0397	0	1	1	10; %FKK400-FKK220
170	155	143	0	0.0198	0	1.05	1	10; %SGPG400-SGPG220
171	115	26	0	0.0208	0	1	2	10; %JRT220-JRT132 tap chng ori 0.95
172	116	27	0	0.0313	0	0.95	2	10; %ARAM220-ARAM132
173	117	28	0	0.0625	0	1	2	10; %KTPS220-KTPS132
174	120	29	0	0.0667	0	1	2	10; %MALPG220-MALPG132
175	122	30	0	0.0208	0	1	2	10; %DRG220-DRG132
176	123	31	0	0.0313	0	1	2	10; %DOMJ220-DOMJ132
177	124	32	0	0.0313	0	1	2	10; %GOK220-GOK132 tap chng
178	125	33	0	0.0625	0	1	2	10; %HOW220-HOW132
179	125	33	0	0.0333	0	1	2	10; %HOW220-HOW132
180	126	34	0	0.0333	0	1	2	10; %KASBA220-KASBA132
181	126	34	0	0.0625	0	1	2	10; %KASBA220-KASBA132
182	127	35	0	0.0313	0	1	2	10; %LXP220-LXP132
183	128	36	0	0.0313	0	0.95	2	10; %MIDNA220-MIDNA132
184	129	37	0	0.0313	0	1	2	10; %NJP220-NJP132
185	130	38	0	0.0313	0	1	2	10; %NHAL220-NHAL132
186	131	39	0	0.0313	0	1	2	10; %RISH220-RISH132
187	132	40	0	0.0313	0	1	2	10; %SATG220-SATG132
188	133	41	0	0.1	0	1	2	10; %DPL220-DPL132

(Continued)

Bus No.	fb	eb	r	x	hlch	Tap	tap posl	lmt
189	136	42	0	0.1	0	1	2	10; %BRPPG220-BRPPG132
190	137	43	0	0.1	0	1	2	10; %SLGPG220-SLGPG132
191	140	95	0	0.1	0	1	2	10; %STPS220-STPS132
192	117	28	0	0.0333	0	1	2	10; %KTPS220-KTPS132
193	136	42	0	0.2	0	1	2	10; %BRPPG220-BRPPG132
194	134	112	0	0.0313	0	1	2	10; %WARIA220-WARIA132
195	144	72	0	0.0313	0	1	2	10; %KRIS220-KRISH132
196	111	164	0	0.2	0	1.05	2	10; %MELLI132-MELLI66 tap chng orig. -1.05
197	10	96	0	0.0324	0	1	2	10; %BTP11-BTPS132 tap chng original -1
198	11	96	0	0.02	0	1	2	10; %BTPS11-BTPS132 tap chng original -1
199	7	140	0	0.0179	0	1	2	10; %STP13.8-STPS220 tap chng original -1
200	8	133	0	0.0769	0	1	2	10; %DPL13.8-DPL220 tap chng original -1
201	12	41	0	0.0314	0	1	2	10; %DPL11-DPL132 tap chng original -1
202	24	41	0	0.1143	0	1	2	10; %DPL6.3-DPL132 tap chng original -1
203	2	117	0	0.0135	0	1	2	10; %KT115.75-KTPS220 tap chng original1
204	3	147	0	0.0169	0	1	2	10; %KT215.75-KTPS400 tap chng originl-1
205	4	148	0	0.0253	0	1	2	10; %BKR15.75-BAKR400 tap chng originl-1
206	1	149	0	0.0112	0	0.95	2	10; %FKK21-FKK400 tap chng original-1
207	5	149	0	0.0177	0	0.95	2	10; %FKK15.75-FKK400 tap chng original-1
208	13	135	0	0.0268	0	0.95	2	10; %CHU11-CHUK220 tap chng original-1
209	9	101	0	0.1333	0	1	2	10; %RMM211-RMM2132 tap chng original-1

(Continued)

Bus No.	fb	eb	r	x	hlch	Tap	tap posl		lmt
210	21	98	0	0.25	0	1	2	10;	%TCF16.6-TCFI132 tap chng original-1
211	22	99	0	0.25	0	1	2	10;	%TCF26.6TCFII132 tap chng original-1
212	23	100	0	0.25	0	1	2	10;	%TCF36.6-TCF3132 tap chng original-1
213	14	158	0	0.1571	0	1	2	10;	%JAL111-JALI66 tap chng original-1
214	15	159	0	0.3125	0	1	2	10;	%JAL211-JALII66 tap chng original-1
215	16	104	0	0.0889	0	1	2	10;	%RANGIT11-RANGT132 tap chng orig.-1
216	53	162	0	0.1333	0	1.05	2	10;	%BRP66-BRP132 tap chng
217	17	141	0	0.0101	0	1	2	10;	%MEJIA11-141MEJIA220
218	6	118	0	0.0405	0	1	2	10;	%BAKR15.7-BAKR220
219	18	112	0	0.0485	0	1	2	10;	%WARIA11-WARIA132
220	19	112	0	0.028	0	1	2	10;	%WARIA21-WARIA132
221	25	157	0	0.0112	0	0.9875	2	10;	%PPSP16.5-PPSP2 tap chng orig.-0.9875
222	39	194	0	0.0533	0	1.05	1	10;	%RISH132-RISHRA33 tap chng orig.-1.05
223	36	201	0	0.127	0	1.05	1	10;	%MIDNA132-MIDNAP33 tap chng orig.-1.05
224	54	195	0	0.08	0	1.05	1	10;	%VISH132-BISHNU33 tap chng orig.-1.05
225	64	196	0	0.254	0	1.05	1	10;	%EGRA132-EGRA33 tap chng original-1.05
226	64	196	0	0.4	0	1.05	1	10;	%EGRA132-EGRA33 tap chng original-1.05
227	77	197	0	0.127	0	1.05	1	10;	%NBU132-NBU33 tap chng original-1.05
228	86	198	0	0.0533	0	1.05	1	10;	%SL132-SLAKE33 tap chng original-1.05

(Continued)

Bus No.	fb	eb	r	x	hlch	Tap	tap posl	lmt
229	55	199	0	0.16	0	1.05	1	10; %BOLP132-BOLPUR33 tap chng orig.-1.05
230	55	199	0	0.2	0	1.05	1	10; %BOLP132-BOLPUR33 tap chng orig.-1.05
231	92	200	0	0.08	0	1.05	1	10; %TITA132-TITAG33 Tap chng original-1.05
232	20	156	0	0.0156	0	0.9875	2	10; %TALA11-TALA400 tap chng orig.-0.9875
233	145	165	0	0.3333	0	1	1	10; %JRT400-220-JRT33 tap chng original-1
234	146	166	0	0.3333	0	1	1	10; %ARAM400-220-ARAM33 tap chng original-1
235	147	167	0	0.5	0	1	1	10; %KTPS400-220-KTPS33 tap chng original-1
236	148	170	0	0.5	0	1	1	10; %BAKR400-220-BAKR33 Tap chng original-1
237	150	169	0	0.5	0	1	1	10; %PARU400-220-PARU33
238	151	171	0	0.5	0	1	1	10; %MALPG400-220-MLDAPG33
239	152	172	0	0.5	0	1	1	10; %BGURI400-220-BGURI33 tap chng ori-1
240	154	173	0	0.5	0	1	1	10; %PURN400-220-PURN33
241	149	168	0	0.5	0	1	1	10; %FKK400-220-FKK33*trf tap has been chng
242	155	202	0	0.5	0	1	1	10; %SGPG400-220-SGPG33
243	115	174	0	0.0741	0	1	1	10; %115JRT220-132-JRT133
244	116	175	0	0.1111	0	1	1	10; %116ARAM220-132-ARAM133
245	117	176	0	0.2222	0	1	1	10; %117KTPS220-132-KTPS133
246	120	177	0	0.2222	0	1	1	10; %120MALPG220-132-MLDPG133

(Continued)

Bus No.	fb	eb	r	x	hlch	Tap	tap posl		lmt
247	122	178	0	0.0741	0	1	1	10;	%122DRG220-132-DRG33
248	123	179	0	0.1111	0	1	1	10;	%123DOMJ220-132-DOMJ33
249	124	180	0	0.1111	0	1	1	10;	%124GOK220-132-GOK33
250	125	181	0	0.2222	0	1	1	10;	%125HOW220-132-HOW33
251	125	181	0	0.1111	0	1	1	10;	%125HOW220-132-HOW33
252	126	182	0	0.1111	0	1	1	10;	%126KASBA220-132-KASBA33
253	126	182	0	0.2222	0	1	1	10;	%126KASBA220-132-KASBA33
254	127	183	0	0.1111	0	1	1	10;	%127LXP220-LXP33
255	128	184	0	0.1111	0	1	1	10;	%128MIDNA220-132-MIDNA33
256	129	185	0	0.1111	0	1	1	10;	%129NJP220-132-NJP33
257	130	186	0	0.1111	0	1	1	10;	%130NHAL220-132-NHAL33
258	131	187	0	0.1111	0	1	1	10;	%131RISH220-132-RISH33
259	132	188	0	0.1111	0	1	1	10;	%132SATG220-132-SATG33
260	133	189	0	0.3333	0	1	1	10;	%133DPL220-132-DPL33
261	136	190	0	0.3333	0	1	1	10;	%136BRPPG220-132-BRPPG33
262	137	191	0	0.3333	0	1	1	10;	%137SLGPG220-132-SLGPG33
263	140	192	0	0.3333	0	1	1	10;	%140STPS220-132-STPS33
264	117	176	0	0.1111	0	1	1	10;	%117KTPS220-132-KTPS133
265	136	190	0	0.6667	0	1	1	10;	%136BRPPG220-132-BRPPG33
266	134	193	0	0.1111	0	1	1	10;	%134WARIA220-132-WARIA33
267	144	203	0	0.1111	0	1	1	10;]	%144KRIS220-132-KRIS33
%47	129	121	0	0	0	0.0004	1		10;

%%total line = 160 + 35(line trf-3 wdg) + 37 (load trf-2wdg) + 35 (for tertiary dg) = 267 %but here total line 266, line 47 has removed due to it zero imp %Lines are arrange acc to bus voltage level.

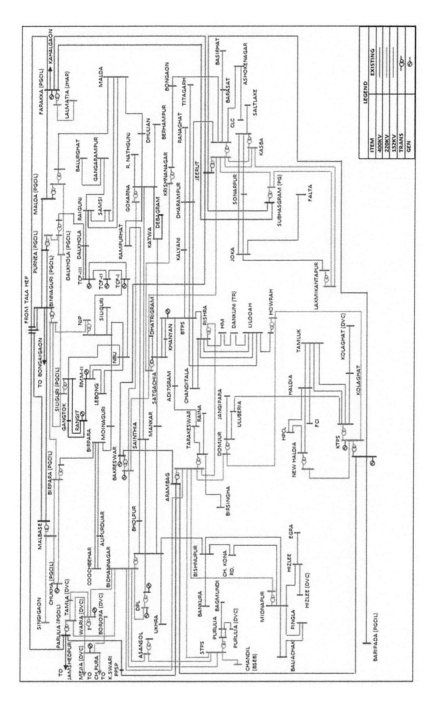

FIGURE B.1 WBSEB's 203-bus system.

Index